A Pedestrian Approach to Quantum Field Theory

A PEDESTRIAN APPROACH TO QUANTUM FIELD THEORY

EDWARD G. HARRIS
University of Tennessee

WILEY-INTERSCIENCE, a Division of John Wiley & Sons, Inc.
New York . London . Sydney . Toronto

Library of Congress Catalog Card Number: 70-37646

ISBN 0-471-35320-5

Printed in the United States of America.

10 9 8 7 6 5 4 3 2 1

The trouble with this world is there are too many metaphysicians that don't know how to tangibilitate.
Father Divine

Preface

For many years it has been customary for all graduate students in physics at the University of Tennessee to take a one-year course in quantum mechanics. As it is now taught, the third quarter of this course is devoted to relativistic wave equations and field theory. No textbook seemed suitable for a one-quarter course in field theory for students of diverse interest, few of whom planned to become theoretical physicists. I therefore prepared my own notes for the course. These changed from year to year, but ultimately settled down enough so that they could be typed and distributed to the students. It then occurred to me that others confronted with the problem of introducing students to field theory in a brief period of time could find these notes useful. With this in mind the notes were expanded and rewritten in book form.

In rewriting the notes I found it advisable to add an introductory chapter on the formalism of quantum mechanics. This contains material that I present in the first quarter of our quantum mechanics course. The well prepared student may find it sufficient to skim through this chapter to acquaint himself with the language and notation that is used. It should serve to introduce the less well prepared student to certain concepts used throughout the book. It is not intended to be an adequate introduction to quantum mechanics for the student with no previous acquaintance with the subject.

It seemed to me to be pedagogically sound to introduce difficult concepts gradually and to apply the theory to physically interesting problems at an early stage of the development of the theory. Therefore in Chapters 2 and 3 we quantize the transverse part of the electromagnetic field, define an interaction Hamiltonian with nonrelativistic charged particles, and apply the theory to some elementary processes in which photons interact with matter. In Chapter 2 I include a section on Glauber's theory of coherent states of the field. Because it is relatively new it does not appear in the standard textbooks on quantum electrodynamics. I include it because of its simplicity and because it clarifies the relation between the classical and quantum-mechanical theories of the field. One of the applications treated in Chapter 2 is the quantum theory of Čerenkov radiation. This phenomenon is interesting and important, and it is also quite simple, since it is a first order process and involves only free particle states. Čerenkov radiation is treated again in Chapter 6. The notion that a particle moving faster than a some wave can emit the wave has important applications in such fields as superfluidity and plasma physics; it therefore seemed desirable to introduce it early in the book.

Having seen how photons emerge from the quantization of the electromagnetic field, the student is prepared to consider the idea that every particle is the quanta of some field. This idea is explored in Chapter 4 where the nonrelativistic Schrödinger equation is quantized. There it is shown that the familiar elementary quantum mechanics is contained in this quantized field theory, but there is more to it than that; there is the possibility of the creation and destruction of particles by the interaction of fields. In Chapter 5 I discuss the interaction of quantized particle fields with the quantized electromagnetic field. Nonrelativistic bremsstrahlung is treated as an example. Finally, in Chapter 6 I discuss quantum electrodynamics in all of its glory. In accordance with the modest aims of this book this discussion is necessarily brief and incomplete. Some tedious calculations have been relegated to an appendix or omitted entirely. All the discussion of infinities and renormalization has been postponed until Chapter 10.

After quantum electrodynamics, the most successful application of quantum field theory has been the theory of beta decay. This theory is briefly discussed in Chapter 7 as an interesting and important application of the ideas of the preceding sections.

In recent years quantum field theory has found important applications in theories of the solid state, plasmas, and liquid helium. An introduction to these applications is given in Chapters 8 and 9.

For all of its many successes quantum field theory contains grave difficulties connected with the divergent integrals that appear in many calculations. I scrupulously avoid these until Chapter 10, where they are finally discussed. I try to give the reader some idea of how the infinite quantities are disposed of in quantum electrodynamics by absorbing them into the mass and charge of the particle—a process known as renormalization. In calculating the Lamb shift and the anomalous magnetic moment of the electron I follow the nonrelativistic theory of Bethe rather than the more exact relativistic theory. This avoids some computational difficulties but serves to introduce the ideas of renormalization.

To make the book self-contained, an appendix on relativistic wave equations is added. All references and some notes concerning these are collected at the end of the book.

The final form of the book contains considerably more material than the lecture notes with which I started. I tried to include a variety of topics in order to give the instructor and students some freedom of choice.

A number of problems are scattered throughout the text. These are intended to supplement the material in the text and to give the student an opportunity to test his understanding. The difficulty of these problems ranges from fairly trivial to fairly difficult. Answers and some solutions are given in an appendix.

I am grateful to Dr. Alvin H. Nielsen, Dean of Liberal Arts, and Dr. William M. Bugg, Head of the Department of Physics, for their very real encouragement in the form of a reduced teaching load which made the writing of this book possible. Many of my colleagues have encouraged me by their interest and suggestions. I am particularly grateful to my quantum mechanics students of this and previous years who have cheerfully endured my experiments in presenting this subject. I also thank Mrs. Patty Martin, Mrs. Wylene Quinn, Mrs. Janice Hemsley, and Miss Jane Pearson for typing the manuscript. Finally, I owe a real debt of gratitude to my wife and daughter for their patience and understanding during the writing of this book.

EDWARD G. HARRIS

Knoxville, Tennessee
October 1971

Contents

A Pedestrian Approach to Quantum Field Theory

1

The Formalism of Quantum Mechanics

It is not an easy task to state the "rules" of quantum mechanics. Many textbooks do not even try and yet succeed in conveying to the reader a working knowledge of the subject. In this book the rules of quantum mechanics and some elementary results are collected in one place for ease of reference. In the sections that follow we give a brief account of the foundations of quantum mechanics. A more detailed discussion of the subject can be found in von Neumann[5] and in the more recent book by Jauch.[6] We begin by discussing the mathematical structure known as a Hilbert space and then give the rules for describing the real world in terms of this mathematical structure.

HILBERT SPACE

A Hilbert space $\mathfrak{H}$ is an abstract set of elements called vectors $|a\rangle$, $|b\rangle$, $|c\rangle$, and so on, having the following set of properties:

1. The space $\mathfrak{H}$ is a linear vector space over the field of complex numbers such as λ, and μ. It has three properties. (*a*) For each pair of vectors there is determined a vector called the sum such that

$$|a\rangle + |b\rangle = |b\rangle + |a\rangle \qquad \text{commutative} \tag{1-1}$$

$$(|a\rangle + |b\rangle) + |c\rangle = |a\rangle + (|b\rangle + |c\rangle) \qquad \text{associative} \tag{1-2}$$

(*b*) One vector $|0\rangle$ is called the null vector.

$$|a\rangle + |0\rangle = |a\rangle \tag{1-3}$$

(*c*) For each vector $|a\rangle$ in $\mathfrak{H}$ there is a vector $|-a\rangle$ such that

$$|a\rangle + |-a\rangle = |0\rangle \tag{1-4}$$

For any complex numbers λ and μ

$$\lambda(|a\rangle + |b\rangle) = \lambda\,|a\rangle + \lambda\,|b\rangle \tag{1-5}$$

$$(\lambda + \mu)\,|a\rangle = \lambda\,|a\rangle + \mu\,|a\rangle \tag{1-6}$$

$$\lambda\mu\,|a\rangle = \lambda(\mu\,|a\rangle) \tag{1-7}$$

$$1\,|a\rangle = |a\rangle \tag{1-8}$$

2. There is defined a scalar product in $\mathfrak{H}$ denoted by $(|a\rangle, |b\rangle)$ or $\langle a \mid b\rangle$. This is a complex number such that

$$(|a\rangle, \lambda\,|b\rangle = \lambda(|a\rangle, |b\rangle \tag{1-9}$$

$$(|a\rangle, |b\rangle + |c\rangle) = (|a\rangle, |b\rangle) + (|a\rangle, |c\rangle) \tag{1-10}$$

$$(|a\rangle, |b\rangle) = (|b\rangle, |a\rangle)^* \tag{1-11}$$

or in the other notation

$$\langle a \mid b\rangle = \langle b \mid a\rangle^* \tag{1-12}$$

It follows that

$$(\lambda\,|f\rangle, |g\rangle) = \lambda^*(|f\rangle, |g\rangle) = \lambda^*\langle f \mid g\rangle \tag{1-13}$$

$$(|f_1\rangle + |f_2\rangle, |g\rangle) = \langle f_1 \mid g\rangle + \langle f_2 \mid g\rangle \tag{1-14}$$

We define the norm of a vector by

$$\text{norm of } |f\rangle = \||f\rangle\| = \sqrt{\langle f \mid f\rangle} \tag{1-15}$$

The following inequality, known as Schwarz's inequality, can be proved:

$$\||f\rangle\| \cdot \||g\rangle\| \leq |\langle f \mid g\rangle| \tag{1-16}$$

The equality sign holds only when $|f\rangle = \lambda\,|g\rangle$.

3. The space $\mathfrak{H}$ is "separable." This means that there exists a sequence $|f_n\rangle$ in $\mathfrak{H}$ with the property that it is dense in $\mathfrak{H}$ in the following sense: for any $|f\rangle$ in $\mathfrak{H}$ and any $\varepsilon > 0$ there exists at least one element $|f_n\rangle$ of the sequence such that

$$\||f\rangle - |f_n\rangle\| < \varepsilon \tag{1-17}$$

4. The space is "complete." This means that any sequence $|f_n\rangle$ with the property

$$\lim_{n, m \to \infty} \||f_n\rangle - |f_m\rangle\| = 0 \tag{1-18}$$

(called a Cauchy sequence) defines a unique limit $|f\rangle$ which is in $\mathfrak{H}$ such that

$$\lim_{n \to \infty} \||f\rangle - |f_n\rangle\| = 0 \tag{1-19}$$

If the vector space has a finite number of dimensions, Axioms III and IV are superfluous, since they follow from Axioms I and II. However, they are necessary for the infinite dimensional spaces which are usual in quantum mechanics.

Now we give some definitions. Two vectors $|f\rangle$ and $|g\rangle$ are said to be "orthogonal" if $\langle f | g\rangle = 0$. A set $\{|f_n\rangle\}$ is said to be an "orthonormal system" if

$$\langle f_n | f_m\rangle = \delta_{nm} \tag{1-20}$$

It is said to be a "complete orthonormal system" of $\mathfrak{H}$ if for every $|f\rangle$ in $\mathfrak{H}$ we have

$$|f\rangle = \sum_n \alpha_n |f_n\rangle \tag{1-21}$$

where the α_n's are complex numbers. Then

$$\langle f_m | f\rangle = \sum_n \alpha_n \langle f_m | f_n\rangle = \alpha_m \tag{1-22}$$

and

$$|f\rangle = \sum_n |f_n\rangle\langle f_n | f\rangle \tag{1-23}$$

The complex numbers $\langle f_n | f\rangle$ are called the representatives of $|f\rangle$. If an infinite number of terms is required in the sum in Eq. 1-21, then $\mathfrak{H}$ is "infinite dimensional." This is the usual case in quantum mechanics.

OPERATORS IN HILBERT SPACE

A linear operator A is defined as a mapping of $\mathfrak{H}$ onto itself (or a subset of $\mathfrak{H}$) such that

$$A(\alpha |f\rangle + \beta |g\rangle) = \alpha A |f\rangle + \beta A |g\rangle \tag{1-24}$$

It is said to be bounded if

$$\|A |f\rangle\| \leq C \, \||f\rangle\| \tag{1-25}$$

with C constant for all $|f\rangle$ in $\mathfrak{H}$. A bounded linear operator A is continuous in the sense that if $|f_n\rangle \rightarrow |f\rangle$ then $A |f_n\rangle \rightarrow A |f\rangle$. We say that $A = B$ if $A |f\rangle = B |f\rangle$ for all $|f\rangle$ in $\mathfrak{H}$.

We define

identity operator 1: $\quad 1 |f\rangle = |f\rangle \quad$ (1-26a)

null operator 0: $\quad 0 |f\rangle = |0\rangle \quad$ (1-26b)

sum of A and B: $\quad (A + B) |f\rangle = A |f\rangle + B |f\rangle \quad$ (1-26c)

product of A and B: $\quad AB |f\rangle = A(B |f\rangle) \quad$ (1-26d)

for all $|f\rangle$ in $\mathfrak{H}$. In general $AB \neq BA$. We call $[A, B] = AB - BA$ the commutator of A and B.

The "adjoint" A^+ of a bounded linear operator A is defined to be a bounded linear operator such that

$$(|g\rangle, A\,|f\rangle) = (A^+\,|g\rangle, |f\rangle) \tag{1-27a}$$

for all $|f\rangle$ and $|g\rangle$ in $\mathfrak{H}$. This may also be written as

$$\langle g|\,A\,|f\rangle = \langle f|\,A^+\,|g\rangle^* \tag{1-27b}$$

The adjoint has the properties

$$(\alpha A)^+ = \alpha^* A^+ \tag{1-28a}$$

$$(A + B)^+ = A^+ + B^+ \tag{1-28b}$$

$$(AB)^+ = B^+A^+ \tag{1-28c}$$

$$(A^+)^+ = A \tag{1-28d}$$

An operator A is said to be "Hermitian" if $A = A^+$. Note that this implies that

$$\langle f|\,A\,|f\rangle = \langle f|\,A^+\,|f\rangle^* = \langle f|\,A\,|f\rangle^* = \text{real} \tag{1-29}$$

EIGENVECTORS AND EIGENVALUES

If A is an operator and there exists a vector $|A'\rangle \neq |0\rangle$ such that

$$A\,|A'\rangle = A'\,|A'\rangle \tag{1-30}$$

where A' is a complex number, then we say that $|A'\rangle$ is an "eigenvector" of A corresponding to the "eigenvalue" A'. Hermitian operators have the following properties:

1. The eigenvalues of a Hermitian operator are real.
2. If $|A'\rangle$ and $|A''\rangle$ are two eigenvectors of a Hermitian operator A, and $A' \neq A''$, then $\langle A' \mid A''\rangle = 0$.
3. The eigenvectors of a bounded Hermitian operator after normalization form a denumerably complete orthonormal system. Consequently, its eigenvalues form a discrete set (discrete spectrum).

It follows that an arbitrary vector $|\psi\rangle$ may be written as

$$|\psi\rangle = \sum_{A'} |A'\rangle\langle A' \mid \psi\rangle \tag{1-31}$$

with

$$\langle A' \mid A''\rangle = \delta_{A'A''} \tag{1-32}$$

The scalar product of two vectors is given by

$$\langle \Phi \mid \psi\rangle = \sum_{A'} \langle \Phi \mid A'\rangle\langle A' \mid \psi\rangle \tag{1-33}$$

A useful trick for remembering this is to write the unit operator as

$$1 = \sum_{A'} |A'\rangle\langle A'| \tag{1-34}$$

Then

$$|\psi\rangle = 1\,|\psi\rangle = \sum_{A'} |A'\rangle\langle A' \mid \psi\rangle \tag{1-35}$$

and

$$\langle\Phi \mid \psi\rangle = \langle\Phi|\,1\,|\psi\rangle = \sum_{A'} \langle\Phi \mid A'\rangle\langle A' \mid \psi\rangle \tag{1-36}$$

Now, every quadratically summable function $\langle A' \mid \psi\rangle$ represents a vector in a Hilbert space. The abstract Hilbert space therefore is mapped onto the space of quadratically summable functions on the spectrum of A. We call this the "A-representation." The action of an operator B on $|\psi\rangle$ is represented by

$$(|A'\rangle, B\,|\psi\rangle) = \langle A'|\,B\,|\psi\rangle = \sum_{A''} \langle A'|\,B\,|A''\rangle\langle A'' \mid \psi\rangle \tag{1-37}$$

In the A-representation a vector $|\psi\rangle$ is represented by the set of complex numbers $\langle A' \mid \psi\rangle$ which may be arranged into a column vector. The operator B is represented by the set of complex numbers $\langle A'|\,B\,|A''\rangle$ which may be arranged into a matrix. For brevity we sometimes write

$$|\psi\rangle \quad \text{or} \quad \langle A' \mid \psi\rangle \quad = \begin{bmatrix} \langle A_1 \mid \psi\rangle \\ \langle A_2 \mid \psi\rangle \\ \cdot \\ \cdot \\ \cdot \\ \langle A_n \mid \psi\rangle \\ \cdot \\ \cdot \\ \cdot \end{bmatrix} \tag{1-38a}$$

$$B \quad \text{or} \quad \langle A'|\,B\,|A''\rangle \quad = \begin{bmatrix} \langle A_1|\,B\,|A_1\rangle & \langle A_1|\,B\,|A_2\rangle & \cdots \\ \langle A_2|\,B\,|A_1\rangle & \langle A_2|\,B\,|A_2\rangle & \cdots \\ \cdot & \cdot & \cdot \\ \cdot & \cdot & \cdot \\ \cdot & \cdot & \cdot \end{bmatrix} \tag{1-38b}$$

Note that in the A-representation the operator A is diagonal; that is,

$$\langle A'|\,A\,|A''\rangle = A'\delta_{A'A''} \tag{1-39}$$

It is sometimes convenient to write an operator in the form

$$B = 1\,B\,1 = \sum_{A'}\sum_{A''} |A'\rangle\langle A'|\,B\,|A''\rangle\langle A''| \tag{1-40}$$

The choice of representation constitutes effectively a choice of coordinate system in Hilbert space. One transforms from the A-representation to the B-representation by using the so-called transformation functions $\langle A' \mid B' \rangle$. An easy way to remember how to do this is to use the unit operator in the forms

$$1 = \sum_{A'} |A'\rangle\langle A'| = \sum_{B'} |B'\rangle\langle B'|, \ldots \tag{1-41}$$

Then

$$\langle B' \mid \psi \rangle = \langle B' | \, 1 \, | \psi \rangle = \sum_{A'} \langle B' \mid A' \rangle \langle A' \mid \psi \rangle \tag{1-42a}$$

$$\langle A' \mid \psi \rangle = \langle A' | \, 1 \, | \psi \rangle = \sum_{B'} \langle A' \mid B' \rangle \langle B' \mid \psi \rangle \tag{1-42b}$$

$$\langle B' | \, C \, | B'' \rangle = \langle B' | \, 1 \, C \, 1 \, | B'' \rangle = \sum_{A'} \sum_{A''} \langle B' \mid A' \rangle \langle A' | \, C \, | A'' \rangle \langle A'' \mid B'' \rangle \tag{1-42c}$$

Note that the product of two operators has the matrix element

$$\langle A' | \, BC \, | A'' \rangle = \langle A' | \, B \, 1 \, C \, | A'' \rangle = \sum_{A'''} \langle A' | \, B \, | A''' \rangle \langle A''' | \, C \, | A'' \rangle \tag{1-43}$$

This is just the rule for multiplying matrices.

Problem 1-1. Show that the trace of an operator is independent of representation, that is,

$$\operatorname{Tr} C = \sum_{A'} \langle A' | \, C \, | A' \rangle = \sum_{B'} \langle B' | \, C \, | B' \rangle = \sum_{C'} C' \tag{1-44}$$

Problem 1-2. Show that

$$\sum_{A'} \sum_{A''} |\langle A' | \, C \, | A'' \rangle|^2 = \operatorname{Tr} CC^+ \tag{1-45}$$

In quantum mechanics we sometimes must consider representations corresponding to operators that have continuous rather than discrete eigenvalues. This causes some difficulties, since there are no proper eigenvectors corresponding to the continuous spectrum. However, we can formally proceed using improper eigenvectors and replacing sums by integrals. Thus

$$|\psi\rangle = \int |A'\rangle \langle A' \mid \psi \rangle \, dA' \tag{1-46}$$

replaces Eq. 1-35. The orthonormality condition, Eq. 1-32, is replaced by

$$\langle A' \mid A'' \rangle = \delta(A' - A'') \tag{1-47}$$

The Dirac δ-function replaces the Kronecker-δ.

In the case of continuous spectra we often write $\langle A' \mid \psi \rangle$ as $\psi(A')$ which we may call the "wave function" in A'-space. The scalar product of two vectors becomes

$$\langle \psi \mid \Phi \rangle = \int \langle \psi \mid A' \rangle \langle A' \mid \Phi \rangle \, dA' = \int \psi^*(A') \Phi(A') \, dA' \tag{1-48}$$

Some operators have mixed spectra. The Hamiltonian for the hydrogen atom is an example. Its eigenvalues are discrete for bound states and continuous for unbound states. In such cases we write

$$|\psi\rangle = \sum_{A'} |A'\rangle\langle A' \mid \psi\rangle + \int |A'\rangle\langle A' \mid \psi\rangle \, dA' \tag{1-49}$$

We can make our notation more compact if we agree to let either $\sum_{A'}$ or $\int dA'$ denote a sum over the discrete part of the spectrum (if any) and an integral over the continuous part (if any).

Functions of operators can be defined in terms of the power series for the function if one exists; that is, if

$$f(x) = \sum_{n=0}^{\infty} C_n x^n \tag{1-50a}$$

then

$$f(A) = \sum_{n=0}^{\infty} C_n A^n \tag{1-50b}$$

defines the function $f(A)$ of the operator A. In this way we may define e^A, $\sin A$, and so on.

Another way of defining $f(A)$ is by means of the eigenvalues. If $A\,|A'\rangle = A'\,|A'\rangle$ then $f(A)\,|A'\rangle = f(A')\,|A'\rangle$.

Problem 1-3. Show that

$$\langle B'|\, f(A)\, |B''\rangle = \sum_{A'} \langle B' \mid A'\rangle f(A')\langle A' \mid B''\rangle \tag{1-51}$$

Problem 1-4. Let σ_x be the 2×2 matrix

$$\sigma_x = \begin{pmatrix} 0 & 1 \\ 1 & 0 \end{pmatrix} \tag{1-52}$$

Show by the power series method and also by using Eq. 1-51 that

$$e^{i(\beta/2)\sigma_x} = \begin{bmatrix} \cos \beta/2 & i \sin \beta/2 \\ i \sin \beta/2 & \cos \beta/2 \end{bmatrix} \tag{1-53}$$

The inverse of an operator can be defined by

$$A^{-1}\,|A'\rangle = \frac{1}{A'}\,|A'\rangle \tag{1-54}$$

Then $A^{-1}A = AA^{-1} = 1$. The inverse does not exist if any A' vanishes.

UNITARY TRANSFORMATIONS

An operator U is called unitary if $U^{-1} = U^+$. Consider a so-called unitary transformation in which vectors are transformed as

$$|A'\rangle_{\text{new}} = U\,|A'\rangle_{\text{old}} \tag{1-55a}$$

and operators are transformed as

$$A_{\text{new}} = UA_{\text{old}}U^+ \tag{1-55b}$$

Then

$$_{\text{new}}\langle B' \mid A'\rangle_{\text{new}} = {}_{\text{old}}\langle B'|\, U^+U\,|A'\rangle_{\text{old}} = {}_{\text{old}}\langle B' \mid A'\rangle_{\text{old}} \tag{1-56}$$

so that scalar products are invariant under a unitary transformation. Also

$$A_{\text{new}}\,|A'\rangle_{\text{new}} = UA_{\text{old}}U^+U\,|A'\rangle_{\text{old}} = A'\,|A'\rangle_{\text{new}} \tag{1-57}$$

so that the eigenvalues A_{new} are the same as those of A_{old}. Furthermore, if

$$C_{\text{old}} = A_{\text{old}}B_{\text{old}} \tag{1-57a}$$

and

$$D_{\text{old}} = A_{\text{old}} + B_{\text{old}} \tag{1-57b}$$

then it is easy to show that

$$C_{\text{new}} = A_{\text{new}}B_{\text{new}} \tag{1-57c}$$

and

$$D_{\text{new}} = A_{\text{new}} + B_{\text{new}} \tag{1-57d}$$

It is straightforward to generalize this to show that all algebraic relations are preserved by unitary transformations.

DIRECT PRODUCT SPACE

It is sometimes desirable to expand the Hilbert space by a process known as the direct product. This is most easily made clear by an example.

A nucleon may be either a proton or a neutron. It is convenient to consider these as two states of the same particle which may be represented by vectors in charge (or isotopic spin) space. We let

$$|p\rangle = \begin{pmatrix}1\\0\end{pmatrix}_{\text{charge}}, \qquad |n\rangle = \begin{pmatrix}0\\1\end{pmatrix}_{\text{charge}} \tag{1-58}$$

These vectors span the two-dimensional charge space. Now, a nucleon can have its spin up or down. We let

$$|\uparrow\rangle = \begin{pmatrix}1\\0\end{pmatrix}_{\text{spin}}, \qquad |\downarrow\rangle = \begin{pmatrix}0\\1\end{pmatrix}_{\text{spin}} \tag{1-59}$$

be the two vectors that span the spin space of the nucleon. The direct product space is the four-dimensional space spanned by the vectors:

$$|p\uparrow\rangle = \begin{pmatrix}1\\0\end{pmatrix}_{\text{charge}} \times \begin{pmatrix}1\\0\end{pmatrix}_{\text{spin}} = \begin{pmatrix}1\\0\\0\\0\end{pmatrix} \tag{1-60a}$$

$$|p\downarrow\rangle = \begin{pmatrix}1\\0\end{pmatrix}_{\text{charge}} \times \begin{pmatrix}0\\1\end{pmatrix}_{\text{spin}} = \begin{pmatrix}0\\1\\0\\0\end{pmatrix} \tag{1-60b}$$

$$|n\uparrow\rangle = \begin{pmatrix}0\\1\end{pmatrix}_{\text{charge}} \times \begin{pmatrix}1\\0\end{pmatrix}_{\text{spin}} = \begin{pmatrix}0\\0\\1\\0\end{pmatrix} \tag{1-60c}$$

$$|n\downarrow\rangle = \begin{pmatrix}0\\1\end{pmatrix}_{\text{charge}} \times \begin{pmatrix}0\\1\end{pmatrix}_{\text{spin}} = \begin{pmatrix}0\\0\\0\\1\end{pmatrix} \tag{1-60d}$$

This direct product space is large enough to accommodate both the spin and the charge attributes of the nucleon. If one desires to accommodate still other attributes, the space must be expanded.

THE AXIOMS OF QUANTUM MECHANICS

We assume the following correspondence between physical quantities and the mathematical objects defined in earlier sections:

1. The state of a physical system corresponds to a ray vector in a Hilbert space $\mathfrak{H}$. This means that $|\psi\rangle$ and $\lambda\,|\psi\rangle$ represent the same state. We shall generally assume the state vectors to be normalized to unity.

2. The dynamical observables of a physical system correspond to "observable operators" in $\mathfrak{H}$. By observable operator we mean a Hermitian operator whose eigenvectors form a basis in which any vector of $\mathfrak{H}$ can be expanded.

We now state some basic physical axioms.

AXIOM I. The result of any measurement of an observable can only be one of the eigenvalues of the corresponding operator. As a result of the measurement the system finds itself in the state represented by the corresponding eigenvector.

AXIOM II. If a system is known to be in the state $|A'\rangle$, then the probability that a measurement of B yields the value B' is

$$W(A', B') = |\langle A' \mid B'\rangle|^2 \tag{1-61a}$$

If B has a continuous spectrum, then

$$|\langle A' \mid B'\rangle|^2 \, dB' \tag{1-61b}$$

is the probability of B having a value in the range B' to $B' + dB'$.

AXIOM III. The operators A and B corresponding to the classical dynamical variables A and B satisfy the following commutation relation:

$$[A, B] = AB - BA = i\hbar\{A, B\}_{\text{op}} \tag{1-62}$$

where $\{A, B\}_{\text{op}}$ is the operator corresponding to the classical Poisson bracket

$$\{A, B\} = \sum_i \left\{\frac{\partial A}{\partial q_i}\frac{\partial B}{\partial p_i} - \frac{\partial A}{\partial p_i}\frac{\partial B}{\partial q_i}\right\} \tag{1-63}$$

and q_i and p_i are the classical coordinates and momenta of the system. One easily finds from this that

$$[q_i, q_j] = [p_i, p_j] = 0 \tag{1-64a}$$

$$[q_i, p_j] = i\hbar\, \delta_{ij} 1 \tag{1-64b}$$

Problem 1-5. The orbital angular momentum is given by $\mathbf{L} = \mathbf{x} \times \mathbf{p}$. Show that

$$[L_x, L_y] = i\hbar L_z \tag{1-65a}$$

This can be generalized to

$$\mathbf{L} \times \mathbf{L} = i\hbar\mathbf{L} \tag{1-65b}$$

One consequence of this axiom deserves mention before we proceed. If we define the expectation value of an observable by

$$\langle A\rangle = \langle\psi|\, A\, |\psi\rangle \tag{1-66}$$

and the uncertainty by

$$\Delta A = \langle(A - 1\langle A\rangle)^2\rangle^{1/2} \tag{1-67}$$

then it can be shown that

$$(\Delta A^2)(\Delta B)^2 \geq -1/4\langle[A, B]\rangle^2 \qquad (1\text{-}68)$$

Applying this to Eq. 1-64b gives the Heisenberg uncertainty relations

$$\Delta p_i \, \Delta q_j \geq \frac{\hbar}{2} \, \delta_{ij} \qquad (1\text{-}69)$$

So far we have been concerned with vectors and observables at one instant of time. The dynamics of a system can be described in several equivalent ways. We discuss first the "Schrödinger-picture" (or representation) in which the state vector is a function of time and the observable operators are time independent.

AXIOM IV. Let the state of the system at the time t_0 be $|\psi_{t_0}\rangle$ and the state of the system at time t be $|\psi_t\rangle$, then the two states are related by the unitary transformation

$$|\psi_t\rangle = U(t - t_0) \, |\psi_{t_0}\rangle \qquad (1\text{-}70)$$

where

$$U(t - t_0) = e^{i\hbar H(t-t_0)} \qquad (1\text{-}71)$$

and H is the Hamiltonian operator of the system. Letting $t - t_0 = dt$, $|\psi_{t_0+dt}\rangle - |\psi_{t_0}\rangle = d\,|\psi\rangle$ and

$$U(dt) = 1 - i/\hbar H \, dt \qquad (1\text{-}72)$$

we find

$$-\frac{\hbar}{i} \frac{\partial}{\partial t} |\psi\rangle = H\psi \qquad (1\text{-}73)$$

This is the Schrödinger equation. (*Note:* in writing Eqs. 1-70 and 1-71 we have assumed that H is independent of time. This is sufficiently general for the purposes of this book. Equation 1-73 is valid even when H is time dependent.)

An equivalent way of describing the dynamics is by the "Heisenberg-picture" (or representation). To accomplish this we let $U = U(t - t_0)$ and consider the unitary transformation

$$|\psi_t\rangle_H = U^{-1} \, |\psi_t\rangle_S = U^{-1}U \, |\psi_{t_0}\rangle_S = |\psi_{t_0}\rangle_S \qquad (1\text{-}74)$$

and

$$A_H(t) = U_t^{-1} A_S U_t \qquad (1\text{-}75)$$

The subscripts S and H stand for Schrödinger-picture and Heisenberg-picture. The operator $|\psi_t\rangle_H = |\psi_{t_0}\rangle_S$ is a fixed vector. The operator

$$A_H(t) = e^{i/\hbar H(t-t_0)} A_S e^{-i/\hbar H(t-t_0)} \qquad (1\text{-}76)$$

is time dependent. Differentiating we find that $A_H(t)$ obeys the equation

$$-\frac{\hbar}{i}\frac{\partial}{\partial t}A_H = A_H H - HA_H = [A_H, H] \tag{1-77}$$

This is the Heisenberg equation of motion for the operator A_H. It may be compared with the classical equation of motion of a dynamical variable in Poisson bracket form

$$\frac{dA}{dt} = \{A, H\} \tag{1-78}$$

We see immediately from Eq. 1-77 that an operator that commutes with the Hamiltonian is a constant of the motion.

A USEFUL THEOREM

Consider two operators A and B which commute; that is,

$$AB = BA \tag{1-79}$$

Let

$$A\,|A'\rangle = A'\,|A'\rangle \tag{1-80}$$

and consider the vector $B\,|A'\rangle$. Operating on $B\,|A'\rangle$ with A and using Eq. 1-79 we find

$$AB\,|A'\rangle = BA\,|A'\rangle = A'B\,|A'\rangle \tag{1-81}$$

We conclude that $B\,|A'\rangle$ is an eigenvector of A corresponding to the eigenvalue A'. If A' is nondegenerate, then $B\,|A'\rangle$ can only differ from $|A'\rangle$ by a constant. Let us call the constant B', then

$$B\,|A'\rangle = B'\,|A'\rangle \tag{1-82}$$

and we see that $|A'\rangle$ is simultaneously an eigenvector of both A and B with eigenvalues A' and B', respectively. We can write it as $|A', B'\rangle$.

In the case of degeneracy this argument must be modified. Suppose that there are a number of vectors $|A', \alpha\rangle$ with $\alpha = 1, 2, \ldots, n$, all of which satisfy

$$A\,|A', \alpha\rangle = A'\,|A', \alpha\rangle \tag{1-83}$$

Then from Eq. 1-81 we can only conclude that $B\,|A'\rangle$ is some linear combination of the vectors $|A', \alpha\rangle$. Often it is desirable to choose the vectors $|A', \alpha\rangle$ so that they are eigenvectors of B. The hydrogen atom problem is an example. There the Hamiltonian H, the square of the orbital angular momentum L^2, and the z-component of the orbital angular momentum L_z all commute with one another. The hydrogen atom wave functions are usually chosen to be eigenfunctions of all three operators although, because of the degeneracy, they need not be.

We now illustrate the general theory of the preceding sections with some simple examples.

SPIN ½ PARTICLE IN A MAGNETIC FIELD

We ignore all of the attributes of the particle except its spin and the magnetic moment associated with it. The angular momentum of a spin ½ particle is given by

$$\mathbf{J} = \frac{\hbar}{2}\sigma \tag{1-84}$$

where

$$\sigma_x = \begin{pmatrix} 0 & 1 \\ 1 & 0 \end{pmatrix}, \qquad \sigma_y = \begin{pmatrix} 0 & -i \\ i & 0 \end{pmatrix}, \qquad \sigma_z = \begin{pmatrix} 1 & 0 \\ 0 & -1 \end{pmatrix} \tag{1-85}$$

are called the Pauli matrices. The energy of a magnetic moment, $\boldsymbol{\mu}$, in a magnetic field $\mathbf{B}$ is given by $-\boldsymbol{\mu}\cdot\mathbf{B}$. We take $\mathbf{B}$ to be in the z-direction and $\boldsymbol{\mu}$ proportional to $\mathbf{J}$. Then with the proper choice of the proportionality constant we can write

$$H = \hbar\omega\sigma_z \tag{1-86}$$

for the Hamiltonian operator. The constant ω has the dimensions of a frequency. The state vectors of this system are vectors in a two-dimensional Hilbert space. This makes the system a particularly simple one to discuss.

First we note that the components of $\mathbf{J}$ do not commute with one another. However, J_z and H do commute, so we can find vectors that are simultaneously eigenvectors of J_z and H. They are readily found to be

$$|\uparrow, z\rangle = \begin{pmatrix} 1 \\ 0 \end{pmatrix} \quad \text{and} \quad |\downarrow, z\rangle = \begin{pmatrix} 0 \\ 1 \end{pmatrix} \tag{1-87}$$

For these vectors

$$J_z\,|\uparrow, z\rangle = \frac{\hbar}{2}\,|\uparrow, z\rangle \tag{1-88a}$$

$$J_z\,|\downarrow, z\rangle = -\frac{\hbar}{2}\,|\downarrow, z\rangle \tag{1-88b}$$

$$H\,|\uparrow, z\rangle = \hbar\omega\,|\uparrow, z\rangle \tag{1-88c}$$

$$H\,|\downarrow, z\rangle = -\hbar\omega\,|\downarrow, z\rangle \tag{1-88d}$$

Now let us consider the operator

$$J_{\mathbf{n}} = \mathbf{J}\cdot\mathbf{n} \tag{1-89a}$$

here **n** is a unit vector

$$\mathbf{n} = \mathbf{e}_x \sin\theta\cos\phi + \mathbf{e}_y \sin\theta\sin\phi + \mathbf{e}_z\cos\theta \tag{1-89b}$$

We find

$$J_{\mathbf{n}} = \frac{\hbar}{2}\begin{pmatrix} \cos\theta & \sin\theta e^{-i\phi} \\ \sin\theta e^{+i\phi} & -\cos\theta \end{pmatrix} \tag{1-90}$$

This is the operator for angular momentum about an axis in the direction of **n**. The eigenvalue problem

$$J_n\,|J_n'\rangle = J_n'\,|J_n'\rangle$$

is readily solved. The eigenvalues are found to be $\pm\hbar/2$. The eigenvectors are

$$|\uparrow, \mathbf{n}\rangle = \begin{pmatrix} \cos\theta/2 \\ \sin\theta/2 e^{i\phi} \end{pmatrix} \tag{1-91a}$$

$$|\downarrow, \mathbf{n}\rangle = \begin{pmatrix} \sin\theta/2 e^{-i\phi} \\ -\cos\theta/2 \end{pmatrix} \tag{1-91b}$$

As $\theta \to 0$ these reduce to $|\uparrow, z\rangle$ and $|\downarrow, z\rangle$ as expected.

Let us suppose that we measure the angular momentum in the direction of **n**. Axiom I tells us that we must find one of the eigenvalues of $J_{\mathbf{n}}$, namely $+\hbar/2$ or $-\hbar/2$. Let us suppose that it is $+\hbar/2$. Then immediately after the measurement we know the state of the system is $|\uparrow, \mathbf{n}\rangle$. Suppose that we now measure J_z while the system is in this state. Axiom II tells us that the probability of finding $+\hbar/2$ is

$$|\langle\uparrow, z \mid \uparrow, \mathbf{n}\rangle|^2 = \cos^2\theta/2 \tag{1-92a}$$

and the probability of finding $-\hbar/2$ is

$$|\langle\downarrow, z \mid \uparrow, \mathbf{n}\rangle|^2 = \sin^2\theta/2 \tag{1-92b}$$

These probabilities add up to unity as they should and have the expected behavior in the limits $\theta \to 0$ and $0 \to \pi$.

Next we discuss the dynamics of the system. We write the state vector as

$$|\psi_t\rangle = \begin{pmatrix} \psi_1(t) \\ \psi_2(t) \end{pmatrix} \tag{1-93}$$

Equations 1-73 and 1-86 give

$$\frac{d}{dt}\begin{pmatrix} \psi_1 \\ \psi_2 \end{pmatrix} = -i\omega\begin{pmatrix} 1 & 0 \\ 0 & -1 \end{pmatrix}\begin{pmatrix} \psi_1 \\ \psi_2 \end{pmatrix} \tag{1-94}$$

from which

$$|\psi_t\rangle = \begin{pmatrix} \psi_1(0)e^{-i\omega t} \\ \psi_2(0)e^{+i\omega t} \end{pmatrix} \tag{1-95}$$

where $\psi_1(0)$ and $\psi_2(0)$ are the initial values of $\psi_1(t)$ and $\psi_2(t)$. Suppose that at $t = 0$ we measure the angular momentum of the system and find that it is $+\hbar/2$ aligned along x-axis. We then know that the initial state of the system is

$$|\uparrow, x\rangle = \begin{pmatrix} 1/\sqrt{2} \\ 1/\sqrt{2} \end{pmatrix} \tag{1-96}$$

(This is obtained from Eq. 1-91a by letting $\theta = \pi/2$ and $\phi = 0$). This tells us that $\psi_1(0) = \psi_2(0) = 1/\sqrt{2}$ and

$$|\psi_t\rangle = \frac{1}{\sqrt{2}}\begin{pmatrix} e^{-i\omega t} \\ e^{+i\omega t} \end{pmatrix} \tag{1-97}$$

Suppose that we now ask for the probability of finding the angular momentum to be $\hbar/2$ aligned along the x-axis at time t. By Axiom II this is

$$|\langle\uparrow, x \mid \psi_t\rangle|^2 = \cos^2 \omega t \tag{1-98a}$$

Similar calculations give

$$|\langle\downarrow, x \mid \psi_t\rangle|^2 = \sin^2 \omega t \tag{1-98b}$$

$$|\langle\uparrow, y \mid \psi_t\rangle|^2 = \cos^2 \left(\omega t + \frac{\pi}{4}\right) \tag{1-98c}$$

$$|\langle\downarrow, y \mid \psi_t\rangle|^2 = \sin^2 \left(\omega t + \frac{\pi}{4}\right) \tag{1-98d}$$

Problem 1-6. Supply the missing steps leading to Eqs. 1-98a, b, and c.

Classically, a spinning rigid body with a magnetic moment would precess about the direction of the magnetic field. One detects a similarity to the classical behavior in Eqs. 1-98.

THE FREE PARTICLE

We begin by considering a free particle moving in one dimension and then later generalize to three dimensions. The dynamical variables are the coordinate x, the momentum p, and the Hamiltonian $p^2/2m$. We can write the eigenvalue equations

$$x\,|x'\rangle = x'\,|x'\rangle \tag{1-99}$$

and

$$p\,|p'\rangle = p'\,|p'\rangle \tag{1-100}$$

We assume that the particle can have any position; thus we assume that x' varies continuously between $-\infty$ and $+\infty$. We make a similar assumption about p'. The normalization of the eigenvectors is

$$\langle x' \mid x'' \rangle = \delta(x' - x'') \tag{1-101}$$

$$\langle p' \mid p'' \rangle = \delta(p' - p'') \tag{1-102}$$

The commutation relation

$$[x, p] = xp - px = i\hbar 1 \tag{1-103}$$

is sufficient to determine the matrix elements of p in the x-representation. Taking matrix elements of Eq. 1-103 gives

$$\begin{aligned}\langle x'| \, xp - px \, |x''\rangle &= \langle x'| \, x1p - p1x \, |x''\rangle \\ &= \int dx'''\{\langle x'| \, x \, |x'''\rangle\{x'''| \, p \, |x''\rangle - \langle x'| \, p \, |x'''\rangle\langle x'''| \, x \, |x''\rangle\} \\ &= i\hbar \, \delta(x' - x'')\end{aligned} \tag{1-104}$$

In deriving Eq. 1-104 we have used

$$1 = \int dx''' \, |x'''\rangle\langle x'''| \tag{1-105}$$

Next we use

$$\langle x'| \, x \, |x''\rangle = x' \, \delta(x' - x'') \tag{1-106}$$

to obtain

$$(x' - x'')\langle x'| \, p \, |x''\rangle = i\hbar \, \delta(x' - x'') \tag{1-107}$$

Using the Dirac δ-function identity

$$x \frac{d}{dx} \delta(x) = -\delta(x)$$

we obtain

$$\langle x'| \, p \, |x''\rangle = \frac{\hbar}{i} \frac{\partial}{\partial x'} \delta(x' - x'') \tag{1-108}$$

Problem 1-7. By a similar calculation show that

$$\langle p'| \, x \, |p''\rangle = -\frac{\hbar}{i} \frac{\partial}{\partial p'} \delta(p' - p'') \tag{1-109}$$

By taking a matrix product we can find $\langle x'| \, p^2 \, |x''\rangle$. Thus

$$\begin{aligned}\langle x'| \, p^2 \, |x''\rangle &= \langle x'| \, p1p \, |x''\rangle \\ &= \int d^3x'''\langle x'| \, p \, |x'''\rangle\langle x'''| \, p \, |x''\rangle \\ &= \left(\frac{\hbar}{i'} \frac{\partial}{\partial x'}\right)^2 \delta(x' - x'')\end{aligned} \tag{1-110}$$

In general

$$\langle x'|\, p^n\, |x''\rangle = \left(\frac{\hbar}{i}\frac{\partial}{\partial x'}\right)^n \delta(x - x'') \tag{1-111a}$$

$$\langle p'|\, x^n\, |x''\rangle = \left(-\frac{\hbar}{i}\frac{\partial}{\partial p'}\right)^n \delta(x' - x'') \tag{1-111b}$$

Next we consider the momentum eigenvalue problem in a coordinate representation. We write

$$p\,|p'\rangle = p'\,|p'\rangle$$

$$\langle x'|\, p\, |p'\rangle = \int dx'' \langle x'|\, p\, |x''\rangle \langle x'' \mid p'\rangle = \frac{\hbar}{i}\frac{\partial}{\partial x'} \langle x' \mid p'\rangle$$

$$= p'\langle x' \mid p'\rangle \tag{1-112}$$

from which

$$\langle x' \mid p'\rangle = \psi_{p'}(x') = \frac{1}{(2\pi\hbar)^{1/2}}\, e^{i/\hbar p'x'} \tag{1-113}$$

The constant of integration is chosen so that

$$\langle p'' \mid p'\rangle = \int dx' \psi_{p''}^*(x')\psi_{p'}(x') = \delta(p' - p'') \tag{1-114}$$

Before discussing the dynamics of a free particle we generalize the result to three dimensions. Since by Eq. 1-64a the coordinates x, y, z commute with one another, we can find a vector $|x', y', z'\rangle$ which for brevity we denote by $|\mathbf{x}'\rangle$ which is simultaneously an eigenvector of x, y, and z with eigenvalues x', y', z', respectively. For brevity we write

$$\mathbf{x}\,|\mathbf{x}'\rangle = \mathbf{x}'\,|\mathbf{x}'\rangle \tag{1-115a}$$

and

$$\langle \mathbf{x}'' \mid \mathbf{x}'\rangle = \delta(\mathbf{x}' - \mathbf{x}'') = \delta(x - x')\,\delta(y - y')\,\delta(z - z') \tag{1-115b}$$

Similarly, p_x, p_y, and p_z commute so we can find a vector $|\mathbf{p}'\rangle$ such that

$$\mathbf{p}\,|\mathbf{p}'\rangle = \mathbf{p}'\,|p'\rangle \tag{1-116a}$$

and

$$\langle \mathbf{p}'' \mid \mathbf{p}'\rangle = \delta(\mathbf{p}' - \mathbf{p}'') \tag{1-116b}$$

We can repeat the argument that led to Eq. 1-108 for p_x, p_y, and p_z obtaining

$$\langle \mathbf{x}'|\, p_x\, |\mathbf{x}''\rangle = \frac{\hbar}{i}\frac{\partial}{\partial x'}\delta(\mathbf{x}' - \mathbf{x}'') \tag{1-117}$$

and two similar relations for the matrix elements of p_y and p_x. These can be condensed into the equation

$$\langle \mathbf{x}'|\, \mathbf{p}\, |\mathbf{x}''\rangle = +\frac{\hbar}{i}\frac{\partial}{\partial \mathbf{x}'}\delta(\mathbf{x}' - \mathbf{x}'') \tag{1-118a}$$

A derivation like that which led to Eq. 1-109 yields

$$\langle \mathbf{p}' | \, \mathbf{x} \, | \mathbf{p}'' \rangle = -\frac{\hbar}{i} \frac{\partial}{\partial \mathbf{p}'} \delta(\mathbf{p}' - \mathbf{p}'') \tag{1-118b}$$

The generalization of Eq. 1-113 gives the momentum eigenfunctions in three dimensions

$$\langle \mathbf{x}' \mid \mathbf{p}' \rangle = \psi_{\mathbf{p}'}(\mathbf{x}') = \frac{1}{(2\pi\hbar)^{3/2}} e^{i/\hbar \mathbf{p}' \cdot \mathbf{x}'} \tag{1-119}$$

The Hamiltonian operator in the x-representation and the p-representation is easily found to be

$$\langle \mathbf{x}' | \, H \, | \mathbf{x}'' \rangle = -\frac{\hbar^2}{2m} \nabla^2 \delta(\mathbf{x}' - \mathbf{x}'') \tag{1-120}$$

and

$$\langle \mathbf{p}' | \, H \, | \mathbf{p}'' \rangle = \frac{1}{2m} p'^2 \delta(\mathbf{p}' - \mathbf{p}'') \tag{1-121}$$

We can use Eqs. 1-70 and 1-71 to find $\psi(x', t) = \langle \mathbf{x}' \mid \psi_t \rangle$ in terms of $\psi(\mathbf{x}', t_0)$. Thus

$$|\psi_t\rangle = e^{-i/\hbar H(t-t_0)} |\psi_{t_0}\rangle \tag{1-122a}$$

$$\begin{aligned} \psi(\mathbf{x}', t) &= \langle \mathbf{x}' | \, e^{-i/\hbar H(t-t_0)} |\psi_{t_0}\rangle \\ &= \int d^3x'' \langle \mathbf{x}' | \, e^{-i/\hbar H(t-t_0)} |\mathbf{x}''\rangle \langle \mathbf{x}'' \mid \psi_{t_0}\rangle \\ &= \int d^3x'' G(\mathbf{x}', t \mid \mathbf{x}'', t_0) \psi(\mathbf{x}'', t_0) \end{aligned} \tag{1-122b}$$

where

$$G(x', t \mid x'', t_0) = \langle \mathbf{x}' | \, e^{-i/\hbar H(t-t_0)} |\mathbf{x}''\rangle \tag{1-122c}$$

is called the propagator. It may be found by operations that by now should be familiar. We write

$$G(\mathbf{x}', t \mid \mathbf{x}'', t_0) = \iint \langle \mathbf{x}' \mid \mathbf{p}' \rangle \, d^3p' \cdot \langle \mathbf{p}' | \, e^{-i/\hbar H(t-t_0)} |\mathbf{p}''\rangle \, d^3p'' \langle \mathbf{p}'' \mid \mathbf{x}'' \rangle \tag{1-123}$$

and use

$$\langle p' | \, e^{-i/\hbar H(t-t_0)} |\mathbf{p}''\rangle = e^{-(i/\hbar)(p'^2/2m)(t-t_0)} \delta(\mathbf{p}' - \mathbf{p}'') \tag{1-124}$$

and Eq. 1-119 to obtain

$$G(\mathbf{x}', t \mid \mathbf{x}'', t_0) = \int \frac{d^3p'}{(2\pi\hbar)^3} e^{i/\hbar[\mathbf{p}' \cdot (\mathbf{x}' - \mathbf{x}'') - (p'^2/2m)(t-t_0)]} \tag{1-125}$$

This integration can be carried out with the result

$$G(\mathbf{x}', t \mid \mathbf{x}'', t_0) = \left(\frac{m}{2\pi i \hbar (t - t_0)} \right)^{3/2} e^{(im/2\hbar)[(\mathbf{x}' - \mathbf{x}'')^2/(t-t_0)]} \tag{1-126}$$

We conclude this section with the remark that we can include spin as an attribute of the free particle by taking the direct product of a vector $|\mathbf{x}\rangle$, $|\mathbf{p}\rangle$, or $|\psi\rangle$ with a spin vector that we denote by $|\sigma\rangle$. For a particle of spin $\frac{1}{2}$, $|\sigma\rangle$ could be either of the vectors of Eq. 1-87. Thus we could write

$$|\psi, \sigma\rangle = |\psi\rangle\,|\sigma\rangle \tag{1-127}$$

and

$$\langle\mathbf{x}\mid \psi, \sigma\rangle = \psi(\mathbf{x})\,|\sigma\rangle = \begin{pmatrix}\psi_1(\mathbf{x})\\ \psi_2(\mathbf{x})\end{pmatrix} \tag{1-128}$$

A particle of spin $\frac{1}{2}$ would be represented by a two-component wave function.

THE ONE-DIMENSIONAL HARMONIC OSCILLATOR

As will be seen in the chapters that follow the harmonic oscillator plays an important role in field theory. Its Hamiltonian may be written as

$$H(x, p) = \frac{1}{2m}\,p^2 + \frac{m\omega^2}{2}\,x^2 \tag{1-129}$$

We would like to solve the energy eigenvalue problem

$$H\,|E\rangle = E\,|E\rangle \tag{1-130}$$

We can do this in several different ways. First, we can use the results of the preceding section to write

$$\langle x'|\,H\,|x''\rangle = \hat{H}\left(x', \frac{\hbar}{i}\frac{\partial}{\partial x'}\right)\,\delta(x' - x'') \tag{1-131a}$$

where

$$\hat{H}\left(x', \frac{\hbar}{i}\frac{\partial}{\partial x'}\right) = -\hbar^2\frac{\partial^2}{\partial x'^2} + \frac{m\omega^2}{2}\,x'^2 \tag{1-131b}$$

Equation 1-130 gives

$$\langle x'|\,H\,|E\rangle = \hat{H}\left(x', \frac{\hbar}{i}\frac{\partial}{\partial x'}\right)\psi_E(x') = E\psi_E(x') \tag{1-132}$$

It is shown in almost all books on quantum mechanics that this differential equation has acceptable solutions only when E has the values

$$E_n = \hbar\omega(n + \tfrac{1}{2}), \qquad n = 0, 1, 2, \ldots, \infty \tag{1-133}$$

These solutions are

$$\psi_n(x') = \left(\frac{m\omega}{\pi\hbar}\right)^{1/4}\frac{1}{\sqrt{2^n n!}}\,H_n(\xi)e^{-\xi^2/2} \tag{1-134a}$$

where

$$\xi = \sqrt{\frac{m\omega}{\hbar}}\, x' \tag{1-134b}$$

and the H_n's are Hermite polynomials.

The same problem can also be solved in the p-representation.

$$\langle p'|\, H \,|p''\rangle = \hat{H}\left(-\frac{\hbar}{i}\frac{\partial}{\partial p'}, p'\right) \delta(p' - p'') \tag{1-135a}$$

where

$$\hat{H}\left(-\frac{\hbar}{i}\frac{\partial}{\partial p'}, p'\right) = \frac{1}{2m} p'^2 - \frac{m\omega^2\hbar^2}{2}\frac{\partial^2}{\partial p'^2} \tag{1-135b}$$

Equation 1-130 gives

$$\hat{H}\left(-\frac{\hbar}{i}\frac{\partial}{\partial p'}, p'\right) \psi_E(p') = E\psi_E(p') \tag{1-136}$$

This equation can be made identical to Eq. 1-132 by an appropriate change of variables.

The probability of finding the particle in the range x' to $x' + dx'$ when its energy is known to be E_n is

$$|\langle x' \mid E_n\rangle|^2\, dx' = |\psi_{E_n}(x')|^2\, dx' \tag{1-137}$$

according to Axiom II. Similarly

$$|\langle p' \mid E_n\rangle|^2\, dp' = |\psi_{E_n}(p')|^2\, dp' \tag{1-138}$$

is the probability of finding the momentum in the range p' to $p' + dp'$.

The coordinate space and momentum space wave functions are related by

$$\langle x' \mid E_n\rangle = \psi_{E_n}(x') = \int dp' \langle x' \mid p'\rangle\langle p' \mid E_n\rangle = \int \frac{dp'}{(2\pi\hbar)^{1/2}}\, e^{i/\hbar \mathbf{p}'\cdot x'} \psi_{E_n}(p') \tag{1-139a}$$

Similarly

$$\psi_{E_n}(p') = \int \frac{dx'}{(2\pi\hbar)^{1/2}}\, e^{-i/\hbar \mathbf{p}'\cdot \mathbf{x}'} \psi_{E_n}(x') \tag{1-139b}$$

Finally, we can solve Eq. 1-130 algebraically without introducing either the x- or p-representations. This will turn out to be the most useful form of the solution for the purposes of this book. We introduce the operators

$$a = \sqrt{\frac{m\omega}{2\hbar}}\, x + \frac{ip}{\sqrt{2m\hbar\omega}} \tag{1-140a}$$

$$a^+ = \sqrt{\frac{m\omega}{2\hbar}}\, x - \frac{ip}{\sqrt{2m\hbar\omega}} \tag{1-140b}$$

Then

$$a^+a = \frac{m\omega}{2\hbar}x^2 + \frac{p^2}{2m\hbar\omega} + \frac{i}{2\hbar}(xp - px) = \frac{1}{\hbar\omega}H - \tfrac{1}{2} \qquad (1\text{-}141)$$

so that

$$H = \hbar\omega(a^+a + \tfrac{1}{2}) = \hbar\omega(N + \tfrac{1}{2}) \qquad (1\text{-}142)$$

where $N = a^+a$. We also find

$$[a, a^+] = \frac{1}{2\hbar}\{i[p, x] - i[x, p]\} = 1 \qquad (1\text{-}143)$$

Denote the eigenvectors of N by $|n\rangle$.

$$N\,|n\rangle = n\,|n\rangle \qquad (1\text{-}144)$$

Now, consider the vector $|b\rangle$ defined by

$$a\,|n\rangle = |b\rangle \qquad (1\text{-}145)$$

Operating on $|b\rangle$ with N we obtain

$$\begin{aligned} N\,|b\rangle &= a^+aa\,|n\rangle = (aa^+ - 1)a\,|n\rangle = (n - 1)a\,|n\rangle \\ &= (n - 1)\,|b\rangle \end{aligned} \qquad (1\text{-}146)$$

We see that $|b\rangle$ is an eigenvector of N with eigenvalue $(n - 1)$. It can only differ from $|n - 1\rangle$ by a constant. We write

$$|b\rangle = a\,|n\rangle = C_n\,|n - 1\rangle \qquad (1\text{-}147)$$

The constant C_n can be evaluated by taking the scalar product of $|b\rangle$ with itself

$$\begin{aligned} (a\,|n\rangle, a\,|n\rangle) &= (C_n\,|n - 1\rangle, C_n\,|n - 1\rangle) = \langle n\,|a^+a\,|n\rangle \\ &= |C_n|^2\,\langle n - 1 \mid n - 1\rangle = n = |C_n|^2 \end{aligned} \qquad (1\text{-}148)$$

Setting an irrelevant phase factor equal to unity, we find $C_n = \sqrt{n}$, and so

$$a\,|n\rangle = \sqrt{n}\,|n - 1\rangle \qquad (1\text{-}149)$$

A similar calculation shows that

$$a^+\,|n\rangle = \sqrt{n + 1}\,|n + 1\rangle \qquad (1\text{-}150)$$

Problem 1-8. Prove Eq. 1-150.

Next we prove that $n \geq 0$. Taking the scalar product of Eq. 1-144 with $|n\rangle$ gives

$$\langle n|\,a^+a\,|n\rangle = n\langle n \mid n\rangle = (a\,|n\rangle, a\,|n\rangle) = n(|n\rangle, |n\rangle) \qquad (1\text{-}151)$$

so that

$$n = \frac{\|a\,|n\rangle\|^2}{\|\,|n\rangle\|} \geq 0 \tag{1-152}$$

Starting with the vector $|n\rangle$ we can generate the sequence $|n-1\rangle$, $|n-2\rangle$, $|n-3\rangle$, and so on, by operating with a. It would seem that the eigenvalue would ultimately become negative which is forbidden. However, if n is an integer the sequence will terminate with $|0\rangle$. We conclude that the eigenvalues of N are the positive integers. It follows that the eigenvalues of H are $\hbar\omega(n + \frac{1}{2})$.

It is useful to have the matrix elements of x and p. Solving Eqs. 1-140 for x and p gives

$$x = \sqrt{\frac{\hbar}{2m\omega}}\,(a^+ + a) \tag{1-153a}$$

$$p = i\sqrt{\frac{m\hbar\omega}{2}}\,(a^+ - a) \tag{1-153b}$$

By using Eqs. 1-149 and 1-150 we immediately find

$$\langle n_1|\, x\, |n_2\rangle = \sqrt{\frac{\hbar}{2m\omega}}\,\{\sqrt{n_2+1}\;\delta_{n_1,n_2+1} + \sqrt{n_2}\;\delta_{n_1,n_2-1}\} \tag{1-154a}$$

$$\langle n_1|\, p\, |n_2\rangle = i\sqrt{\frac{m\hbar\omega}{2}}\,\{\sqrt{n_2+1}\;\delta_{n_1,n_2+1} - \sqrt{n_2}\;\delta_{n_1,n_2-1}\} \tag{1-154b}$$

Problem 1-9. Calculate $\langle n_1|\, x^2\, |n_2\rangle$ and $\langle n_1|\, p^2\, |n_2\rangle$ and use this to show that $\langle n_1|\, H\, |n_2\rangle = \hbar\omega(n_1 + \frac{1}{2})\,\delta_{n_1,n_2}$.

PERTURBATION THEORY

A problem that is often encountered in quantum mechanics is that of finding approximate solutions of

$$-\frac{\hbar}{i}\frac{\partial}{\partial t}\,|\psi\rangle = (H_0 + H')\,|\psi\rangle \tag{1-155}$$

when the solutions of

$$H_0\,|\Phi_n\rangle = E_n\,|\Phi_n\rangle \tag{1-156}$$

are known and H' may in some sense be considered as a small perturbation. If we let

$$|\psi\rangle = \sum_n C_n(t) e^{-i/\hbar E_n t}\,|\Phi_n\rangle \tag{1-157}$$

and use Eq. 1-156, then Eq. 1-155 reduces to the set of coupled differential equations for the coefficients

$$\frac{d}{dt} C_m(t) = -\frac{i}{\hbar} \sum_n \langle \Phi_m | H' | \Phi_n \rangle e^{i/\hbar(E_m - E_n)t} C_n(t) \tag{1-158}$$

By integrating from 0 to t this may be converted to the integral equation

$$C_m(t) = C_m(0) - \frac{i}{\hbar} \sum_n \int_0^t dt' \langle \Phi_m | H' | \Phi_n \rangle e^{i/\hbar(E_m - E_n)t'} C\ (t') \tag{1-159}$$

At this point we introduce an approximation. We assume that at the time $t = 0$ the system is in the state $|\Phi_i\rangle$ so that $C_m(0) = \delta_{mi}$. We assume that because H' is so small none of the C_n's depart appreciably from their initial values. Also we assume that H' is independent of time. Then for $f \neq i$ we find

$$\begin{aligned} C_f(t) &= -\frac{i}{\hbar} \langle \Phi_f | H' | \Phi_i \rangle \int_0^t dt' e^{i/\hbar(E_f - E_i)t'} \\ &= -\frac{i}{\hbar} \langle \Phi_f | H' | \Phi_i \rangle \left[\frac{e^{i/\hbar(E_f - E_i)t} - 1}{i/\hbar(E_f - E_i)} \right] \end{aligned} \tag{1-160}$$

The probability of finding the system in the state $|\Phi_f\rangle$ at time t is $|C_f(t)|^2$. From Eq. 1-160 this is found to be

$$|C_f(t)|^2 = \frac{4}{\hbar^2} |\langle \Phi_f | H' | \Phi_i \rangle|^2 \frac{\sin^2 (\omega_{fi} t/2)}{\omega_{fi}^{\ 2}} \tag{1-161a}$$

where

$$\omega_{fi} = (E_f - E_i)/\hbar \tag{1-161b}$$

Now, regarded as a function of ω, the function $\sin^2 (\omega t/2)/\omega^2$ becomes very sharply peaked about $\omega = 0$ when t becomes large. Most of the area under a graph of the function is under the central peak. Also

$$\int_{-\infty}^{+\infty} d\omega \frac{\sin^2 (\omega t/2)}{\omega^2} = \frac{\pi t}{2} \tag{1-162}$$

Therefore, we can say that

$$\frac{\sin^2 (\omega t/2)}{\omega^2} \xrightarrow[t \to \infty]{} \frac{\pi t}{2} \delta(\omega) \tag{1-163}$$

Using this in Eq. 1-161 gives

$$\frac{|C_f(t)|^2}{t} = \frac{2\pi}{\hbar} |\langle \Phi_f | H' | \Phi_i \rangle|^2 \delta(E_f - E_i) \tag{1-164}$$

This may be interpreted as the transition probability per unit time for a transition from an initial state $|\Phi_i\rangle$ to a final state $|\Phi_f\rangle$. This result is known as "Fermi's golden rule." Since Eq. 1-164 contains a Dirac δ-function, it is clear that it is meaningful only if an integration over a continuum of final energies or initial energies is ultimately carried out.

Higher-order approximations can be found by iterating Eq. 1-159 a number of times. The calculations are tedious and will not be carried out here. However, the results are simple and will be quoted without proof. The transition probability per unit time for the transition $i \rightarrow f$ is given by

$$\left(\frac{\text{trans prob}}{\text{time}}\right)_{i\rightarrow f} = \frac{2\pi}{\hbar} |M_{fi}|^2 \, \delta(E_f - E_i) \tag{1-165}$$

where M_{fi}, the matrix element for the transition, is given by

$$M_{if} = \langle f | \, H' \, | i \rangle + \sum_I \frac{\langle f | \, H' \, | I \rangle \langle I | \, H' \, | i \rangle}{E_i - E_I + i\eta} + \sum_I \sum_{II} \frac{\langle f | \, H' \, | I \rangle \langle I | \, H' \, | II \rangle \langle II | \, H' \, | i \rangle}{(E_i - E_I + i\eta)(E_i - E_{II} + i\eta)} + \cdots \tag{1-166}$$

In this equation we have simplified the notation by using $\langle f | \, H' \, | i \rangle$ for $\langle \Phi_f | \, H' \, | \Phi_i \rangle$, and so on. The states $|I\rangle$, $|II\rangle$, and so on, are intermediate states through which the transition can occur. The quantity η is a positive infinitesimal. It is needed to prescribe how the singularities in the expression for M_{fi} are to be treated.

2

Quantum Theory of the Free Electromagnetic Field

As is well known the electric field **E** and magnetic field **B** can be derived from a scalar potential ϕ and vector potential **A** by the formulas

$$\mathbf{E} = -\frac{1}{c}\frac{\partial \mathrm{A}}{\partial t} - \nabla\phi \tag{2-1a}$$

$$\mathbf{B} = \mathbf{\nabla} \times \mathrm{A} \tag{2-1b}$$

(We use Gaussian units throughout this book.) If there are no sources of the field it is always possible to choose a gauge (called the Coulomb gauge) in which

$$\phi = 0 \tag{2-2a}$$

and

$$\mathbf{\nabla} \cdot \mathbf{A} = 0 \tag{2-2b}$$

Now, consider Maxwell's equations for a field without sources.

$$\mathbf{\nabla} \cdot \mathbf{B} = 0 \tag{2-3a}$$

$$\mathbf{\nabla} \cdot \mathbf{E} = 0 \tag{2-3b}$$

$$\mathbf{\nabla} \times \mathbf{E} = -\frac{1}{c}\frac{\partial \mathbf{B}}{\partial t} \tag{2-3c}$$

$$\mathbf{\nabla} \times \mathbf{B} = \frac{1}{c}\frac{\partial \mathbf{E}}{\partial t} \tag{2-3d}$$

The first three of these equations are satisfied identically when **E** and **B** are given in terms of the potentials by Eqs. 2-1, and ϕ and **A** satisfy Eqs. 2-2. Equation 2-3d gives

$$\nabla^2\mathbf{A} - \frac{1}{c^2}\frac{\partial^2\mathbf{A}}{\partial t} = 0 \tag{2-4}$$

In developing a quantum theory of the electromagnetic field it is convenient to describe the field by a set of discrete variables. To this end we make a Fourier analysis of the field in a large cubical box of volume $\Omega = L^3$ and take the Fourier coefficients as the field variables. The most convenient choice of boundary conditions is to require **A** to be periodic on the walls of the box. Thus we require

$$\mathbf{A}(L, y, z, t) = \mathbf{A}(o, y, z, t) \tag{2-5a}$$

$$\mathbf{A}(x, L, z, t) = \mathbf{A}(x, o, z, t) \tag{2-5b}$$

$$\mathbf{A}(x, y, L, t) = \mathbf{A}(x, y, o, t) \tag{2-5c}$$

We write **A** as the Fourier series

$$\mathrm{A}(x, t) = \sum_{\substack{\mathbf{k} \\ k_z>0}} \sum_{\sigma=1,2} \left(\frac{2\pi hc^2}{\Omega\omega_k}\right)^{1/2} \mathbf{u}_{\mathbf{k}\sigma}\{a_{\mathbf{k}\sigma}(t)e^{i\mathbf{k}\cdot\mathbf{x}} + a^*_{\mathbf{k}\sigma}(t)e^{-i\mathbf{k}\cdot\mathbf{x}}\} \tag{2-6}$$

The factor $(2\pi\hbar c^2/\Omega\omega_k)^{1/2}$ is a normalization factor chosen for later convenience. The vectors $\mathbf{u}_{\mathbf{k}1}$ and $\mathbf{u}_{\mathbf{k}2}$ are two unit polarization vectors; in order that Eq. 2-2b be satisfied they must be chosen perpendicular to **k**. In order for Eqs. 2-5 to be satisfied the wave vectors **k** must have the components $(n_x, n_y, n_z)2\pi/L$ where the n_i are integers. We have written **A** as a complex quantity plus its complex conjugate so as to make **A** real as it should be. Since both $e^{i\mathbf{k}\cdot\mathbf{x}}$ and $e^{-i\mathbf{k}\cdot\mathbf{x}}$ are included in each term of Eq. 2-6 we must restrict the summation to one-half of **k** space; hence the restriction to $k_z > 0$.

Substituting Eq. 2-6 into Eq. 2-4 gives

$$\frac{d^2}{dt^2} a_{\mathbf{k}\sigma} + \omega_k^{\,2} a_{\mathbf{k}\sigma} = 0 \tag{2-7}$$

where $\omega_k = kc$. This has the solution

$$a_{\mathbf{k}\sigma}(t) = a^{(1)}_{\mathbf{k}\sigma}(0)e^{-i\omega_k t} + a^{(2)}_{\mathbf{k}\sigma}(0)e^{+i\omega_k t} \tag{2-8}$$

so we can write

$$\mathbf{A}(\mathbf{x}, t) = \sum_{\substack{\mathbf{k},\sigma \\ k_z>0}} \left(\frac{2\pi\hbar c^2}{\omega_k\Omega}\right)^{1/2} \mathbf{u}_{k\sigma} a^{(1)}_{\mathbf{k}\sigma}(0)e^{i(\mathbf{k}\cdot\mathbf{x}-\omega_k t)}$$

$$+ a^{(1)*}_{\mathbf{k}\sigma}(0)e^{-i(\mathbf{k}\cdot\mathbf{x}-\omega_k t)} + a^{(2)}_{\mathbf{k}\sigma}(0)e^{i(\mathbf{k}\cdot\mathbf{x}+\omega_k t)} + a^{(2)*}_{\mathbf{k}\sigma}(0)e^{-i(\mathbf{k}\cdot\mathbf{x}+\omega_k t)} \tag{2-9}$$

We can get rid of the restriction $k_z > 0$ and simplify this formula by defining

$$a_{\mathbf{k}\sigma}(0) = a^{(1)}_{\mathbf{k}\sigma}(0) \qquad \text{for } k_z > 0 \tag{2-10a}$$

$$a_{\mathbf{k}\sigma}(0) = a^{(2)}_{-\mathbf{k}\sigma}(0) \qquad \text{for } k_z < 0 \tag{2-10b}$$

Equation 2-9 now becomes

$$\mathbf{A}(\mathbf{x}, t) = \sum_{\mathbf{k},\sigma} \left(\frac{2\pi hc^2}{\Omega\omega_k}\right)^{1/2} \mathbf{u}_{\mathbf{k}\sigma}[a_{\mathbf{k}\sigma}(t)e^{i\mathbf{k}\cdot\mathbf{x}} + a^*_{\mathbf{k}\sigma}(t)e^{-i\mathbf{k}\cdot\mathbf{x}}] \tag{2-11a}$$

where

$$a_{\mathbf{k}\sigma}(t) = a_{\mathbf{k}\sigma}(0)e^{-i\omega_k t} \tag{2-11b}$$

so that

$$\frac{d}{dt} a_{\mathbf{k}\sigma} = -i\omega_k a_{\mathbf{k}\sigma} \tag{2-11c}$$

Equations 2-11c for all **k** and σ may be regarded as the equations of motion of the field. We show that they can be derived from a Hamiltonian whose value is the total energy of the field.

The energy in the electromagnetic field is

$$H_{\text{rad}} = \frac{1}{8\pi}\int_\Omega d^3x(E^2 + B^2)$$

$$= \frac{1}{8\pi}\int_\Omega d^3x\left\{\frac{1}{c^2}\left|\frac{\partial \mathbf{A}}{\partial t}\right|^2 + |\nabla \times \mathbf{A}|^2\right\} \tag{2-12}$$

Using Eq. 2-11 we find

$$\int_\Omega d^3x \frac{1}{8\pi c^2}\left|\frac{\partial \mathbf{A}}{\partial t}\right|^2$$

$$= -\sum_{\mathbf{k},\sigma}\sum_{\mathbf{k}',\sigma'} \frac{1}{8\pi c^2}\left(\frac{2\pi hc^2}{\Omega}\right)(\omega_k\omega_{k'})^{1/2}\int d^3x[a_{\mathbf{k}\sigma}e^{i\mathbf{k}\cdot\mathbf{x}} - a^*_{\mathbf{k}\sigma}e^{-i\mathbf{k}\cdot\mathbf{x}}]$$

$$\times [a_{\mathbf{k}'\sigma'}e^{i\mathbf{k}\cdot\mathbf{x}} - a^*_{\mathbf{k}'\sigma'}e^{-i\mathbf{k}'\cdot\mathbf{x}}] \tag{2-13}$$

Now we use

$$\int_\Omega d^3x e^{i(\mathbf{k}-\mathbf{k})\cdot\mathbf{x}} = \Omega\delta_{\mathbf{k},\mathbf{k}'} \tag{2-14}$$

to get rid of the integral over d^3x and the sum over $\mathbf{k}'$. We use

$$\mathbf{u}_{\mathbf{k}\sigma} \cdot u_{\mathbf{k}\sigma'} = \delta_{\sigma,\sigma'} \tag{2-15}$$

to get rid of the sum over σ'. We are left with

$$\int_\Omega d^3x \frac{1}{8\pi c^2}\left|\frac{\partial \mathrm{A}}{\partial t}\right|^2$$

$$= \tfrac{1}{4}\sum_{\mathbf{k},\sigma} \hbar\omega_k\{(a_{\mathbf{k}\sigma}a^*_{\mathbf{k}\sigma} + a^*_{\mathbf{k}\sigma}a_{\mathbf{k}\sigma}) - (a_{\mathbf{k}\sigma}a_{-\mathbf{k}\sigma} + a^*_{\mathbf{k}\sigma}a^*_{-\mathbf{k}\sigma})\} \tag{2-16}$$

When we calculate the contribution of the $|\nabla \times \mathbf{A}|^2$ term to H_{rad} we find a result that differs from Eq. 2-11 only in the sign of the $(a_{\mathbf{k}\sigma}a_{-\mathbf{k}\sigma} + a^*_{\mathbf{k}\sigma}a^*_{-\mathbf{k}\sigma})$

term. When the two contributions are added, these terms cancel and the result is

$$H_{\text{rad}} = \tfrac{1}{2} \sum_{\mathbf{k}\sigma} \hbar\omega_k (a_{\mathbf{k}\sigma} a^*_{\mathbf{k}\sigma} + a^*_{\mathbf{k}\sigma} a_{\mathbf{k}\sigma}) \tag{2-17a}$$

$$H_{\text{rad}} \overset{?}{=} \sum_{\mathbf{k}\sigma} \hbar\omega_k a^*_{\mathbf{k}\sigma} a_{\mathbf{k}\sigma} \tag{2-17b}$$

In calculating H_{rad} we have been careful to maintain the order of the factor in products such as $a_{\mathbf{k}\sigma} a^*_{\mathbf{k}\sigma}$ although at this stage we regard them as classical quantities. Later we shall interpret $a_{\mathbf{k}\sigma}$ and $a^*_{\mathbf{k}\sigma}$ as noncommuting operators and the last step in Eq. 2-17 is questionable. That is the reason for the question mark. This question will be discussed later.

Comparing Eq. 2-17 with Eq. 1-142 we see that H_{rad} resembles the Hamiltonian for a collection of harmonic oscillators. We can treat the radiation field quantum mechanically by interpreting $a_{\mathbf{k}\sigma}$ as an operator and $a^*_{\mathbf{k}\sigma}$, which we henceforth denote by $a^+_{\mathbf{k}\sigma}$, as its adjoint. We assume that the variables referring to different oscillators commute, so in analogy with Eq. 1-143 we assume

$$[a_{\mathbf{k}\sigma}, a^+_{\mathbf{k}'\sigma'}] = \delta_{\mathbf{k},\mathbf{k}'}\, \delta_{\sigma,\sigma} \tag{2-18}$$

Assuming

$$H_{\text{rad}} = \sum_{\mathbf{k},\sigma} h\omega_k a^+_{\mathbf{k}\sigma} a_{\mathbf{k}\sigma} \tag{2-19}$$

The Heisenberg equations of motion

$$-\frac{\hbar}{i}\frac{\partial}{\partial t} a_{\mathbf{k}\sigma} = [a_{\mathbf{k}\sigma}, H] \tag{2-20}$$

yield Eq. 2-11c.

There is a question whether we should have retained the zero-point energy of the oscillators and written

$$H_{\text{rad}} = \sum_{\mathbf{k},\sigma} \hbar\omega_k (a^+_{\mathbf{k}\sigma} a_{\mathbf{k}\sigma} + \tfrac{1}{2}) \tag{2-21}$$

If we do, the zero-point energy of the radiation field

$$\sum_{\mathbf{k}\sigma} \hbar\omega_k/2 \tag{2-22}$$

is infinite because there are an infinite number of field oscillators. For most purposes this infinite energy of the vacuum cancels out when any physically meaningful quantity is calculated, so we shall generally assume that H_{rad} is given by Eq. 2-19.

We can write the state vectors for the electromagnetic field as the direct product of the state vectors for each of the field oscillators. Thus

$$|\cdots n_{\mathbf{k}\sigma} \cdots n_{\mathbf{k}'\sigma'} \cdots\rangle = |\cdots\rangle \cdots |n_{\mathbf{k}\sigma}\rangle \cdots |n_{\mathbf{k}',\sigma'}\rangle \cdots \tag{2-23}$$

where

$$a^{+}_{\mathbf{k}\sigma}a_{\mathbf{k}\sigma}\,|n_{\mathbf{k}\sigma}\rangle = n_{\mathbf{k}\sigma}\,|n_{\mathbf{k}\sigma}\rangle \tag{2-24}$$

and $n_{\mathbf{k}\sigma} = 0, 1, 2, 3, \cdots, \infty$. In analogy with Eqs. 1-149 and 1-150 we find

$$a_{\mathbf{k}\sigma}\,|\cdots n_{\mathbf{k}\sigma}\cdots\rangle = \sqrt{n_{\mathbf{k}\sigma}}\,|\cdots n_{\mathbf{k}\sigma} - 1\cdots\rangle \tag{2-25}$$

$$a^{+}_{\mathbf{k}\sigma}\,|\cdots n_{k\sigma}\cdots\rangle = \sqrt{n_{k\sigma} + 1}\,|\cdots n_{\mathbf{k}\sigma} + 1\cdots\rangle \tag{2-26}$$

These relations are a consequence of Eq. 2-18.

The state vectors of Eq. 2-23 are eigenvectors of H_{rad} with eigenvalues

$$E = \sum_{\mathbf{k},\sigma} = \hbar\omega_k n_{\mathbf{k}\sigma} \tag{2-27}$$

It may be shown that the momentum operator of the field, namely

$$\mathbf{P} = \int_\Omega d^3x\,\frac{\mathbf{E}\times\mathbf{B}}{4\pi c} \tag{2-28a}$$

is given by

$$\mathbf{P} = \sum_{\mathbf{k},\sigma} \hbar\mathbf{k}a^{+}_{\mathbf{k}\sigma}a_{\mathbf{k}\sigma} \tag{2-28b}$$

Therefore the state vectors of Eq. 2-23 are also eigenvectors of **P** with eigenvalues

$$\mathbf{P}' = \sum_{\mathbf{k},\sigma} \hbar k n_{\mathbf{k}\sigma} \tag{2-28c}$$

On the basis of the preceding discussion it is natural to suppose that the electromagnetic field consist of photons each of which has the energy $\hbar\omega_k$ and momentum $\hbar\mathbf{k}$; $n_{\mathbf{k}\sigma}$ is the number of photons with momentum $h\mathbf{k}$ and polarization given by the vector $\mathbf{u}_{\mathbf{k}\sigma}$. Since, when the operator $a_{\mathbf{k}\sigma}$ operates on a state vector, it decreases the number of photons by one, it is called an "annihilation" or "destruction" operator. Similarly, $a^{+}_{\mathbf{k}\sigma}$ is called a "creation" operator since it increases $n_{\mathbf{k}\sigma}$ by one when it operates on a state vector.

COHERENT STATES OF THE RADIATION FIELD

Let us consider the electric field due to one term in the expression for **A** given in Eq. 2-11.

$$\mathbf{E}(\mathbf{x}, t) = -\frac{1}{c}\frac{\partial \mathrm{A}}{\partial t} = -i\left(\frac{2\pi\hbar\omega}{\Omega}\right)^{1/2}\mathbf{u}[ae^{i\mathbf{k}\cdot\mathbf{x}} - a^{+}e^{-i\mathbf{k}\cdot\mathbf{x}}] \tag{2-29}$$

where subscripts that are irrelevant have been dropped. When there are n photons in this mode of the field the expectation value of **E** is

$$\langle n|\,\mathbf{E}\,|n\rangle = 0 \tag{2-30}$$

since

$$\langle n| a |n\rangle = \langle n| a^+ |n\rangle = 0 \tag{2-31}$$

On the other hand, the expectation value of the energy density is

$$\langle n| \frac{E^2}{4\pi} |n\rangle = \frac{\hbar\omega}{\Omega} (n + \tfrac{1}{2}) \tag{2-32}$$

Equations 2-30 and 2-32 are what we would expect if there were n photons in the field, but their phases were random so that when we averaged over the phases the average value of **E** vanished.

Glauber[13] has introduced a state of the field in which **E** behaves more like a classical field. It is necessary to introduce some uncertainty into the number of photons present in order to more precisely define the phase. Let c be a complex number and define the state $|c\rangle$ by

$$|c\rangle = \sum_{n=0}^{\infty} b_n |n\rangle \tag{2-33a}$$

where

$$b_n = \frac{c^n e^{-\frac{1}{2}|c|^2}}{\sqrt{n!}} \tag{2-33b}$$

By the usual rules of quantum mechanics

$$|b_n|^2 = \frac{|c|^{2n} e^{-|c|^2}}{n!} \tag{2-34}$$

is the probability of finding n photons in the field. The sum of these probabilities is unity since

$$\sum_{n=0}^{\infty} |b_n|^2 = e^{-|c|^2} \sum_{n=0}^{\infty} \frac{|c|^{2n}}{n!} = e^{-|c|^2} e^{+|c|^2} = 1 \tag{2-35}$$

In this state the expectation value of a is

$$\begin{aligned}\langle c| a |c\rangle &= \sum_{m=0}^{\infty} \sum_{n=0}^{\infty} b_m^* b_n \sqrt{n} \langle m \mid n - 1\rangle \\ &= \sum_{n=0}^{\infty} b_{n-1}^* b_n \sqrt{n} \\ &= e^{-|c|^2} \sum_{n=0}^{\infty} \frac{(c^*)^{n-1} c^n}{\sqrt{(n-1)!}\,\sqrt{n!}} \sqrt{n} \\ &= c e^{-|c|^2} \sum_{n=0}^{\infty} \frac{|c|^{2n}}{n!} = c \end{aligned} \tag{2-36}$$

where we have used Eq. 2-25. In a similar manner we can show that

$$\langle c| a^+ |c\rangle = c^* \tag{2-37}$$

It follows that the expectation value of **E** is

$$\langle c|\, \mathbf{E}\, |c\rangle = -i\left(\frac{2\pi\hbar\omega}{\Omega}\right)^{1/2} \mathbf{u}[ce^{i\mathbf{k}\cdot\mathbf{x}} - c^* e^{-i\mathbf{k}\cdot\mathbf{x}}] \tag{2-38}$$

This is the form we expect for a classical electromagnetic wave. The amplitude of the wave is determined by the modulus of c and the phase is determined by the phase of c. This is the same form as Eq. 2-29 but the operators a and a^+ have been replaced by the complex numbers c and c^+.

Brief calculations like that of Eq. 2-36 show that

$$\langle c|\, a^+ a\, |c\rangle = |c|^2 = \langle n\rangle \tag{2-39a}$$

$$\langle c|\, a\, a^+\, |c\rangle = |c|^2 + 1 = \langle n\rangle + 1 \tag{2-39b}$$

$$\langle c|\, a^+ a a^+ a\, |c\rangle = |c|^4 + |c|^2 = \langle n^2\rangle \tag{2-39c}$$

$$\langle c|\, a^2\, |c\rangle = c^2 \tag{2-39d}$$

$$\langle c|\, a^{+2}\, |c\rangle = c^{*2} \tag{2-39e}$$

Problem 2-1. Prove Eqs. 2-39a through 2-39c.

We may define the uncertainty of the number of photons in the state $|c\rangle$ in analogy with Eq. 1-67 by

$$\begin{aligned} \Delta n &= \langle c|\, (n - \langle n\rangle)^2\, |c\rangle^{1/2} \\ &= (\langle c|\, n^2\, |c\rangle - \langle n\rangle^2)^{1/2} \\ &= (\langle n^2\rangle - \langle n\rangle^2)^{1/2} = \langle n\rangle^{1/2} \end{aligned} \tag{2-40}$$

The relative uncertainty is

$$\frac{\Delta n}{\langle n\rangle} = \frac{1}{\langle n\rangle^{1/2}} \tag{2-41}$$

This becomes very small when the expectation value of the number of photons in this mode becomes very large.

The point of all this is that if there are a large number of photons in the same mode of the field, then the relative uncertainty of that number can be very small, and the expectation value of **E** behaves like a classical field.

Problem 2-2. Show that

$$\langle c|\, E^2\, |c\rangle - |\langle c|\, \mathbf{E}\, |c\rangle|^2 = \frac{2\pi\hbar\omega}{\Omega} \tag{2-42}$$

This vanishes in the classical limit ($\hbar \to 0$).

3

Interaction of Radiation and Matter

Let us consider a collection of particles of masses m_i and charges e_i that interact through forces derivable from the potential $V(\cdots \mathbf{x}_i \cdots \mathbf{x}_j \cdots)$ which includes the Coulomb forces between the particles. For simplicity of notation we refer to this system as an atom although it may be a molecule, a nucleus, or other system. The Hamiltonian of this system may be written as

$$H_{\text{atom}} = \sum_i \frac{1}{2m_i} p_i^2 + V \tag{3-1}$$

Now, we let this system interact with the electromagnetic field discussed in Chapter 2. There is a simple prescription for modifying a Hamiltonian to include the interaction with an electromagnetic field derivable from a vector potential $\mathbf{A}$. The prescription is to replace $\mathbf{p}_i$ by $\mathbf{p}_i - e_{i/c}\mathbf{A}(\mathbf{x}_i)$. If we do this in Eq. 3-1 and add on the Hamiltonian of the radiation field we get the Hamiltonian for the combined system:

$$\begin{aligned} H &= \sum_i \frac{1}{2m_i} |\mathbf{p}_i - \frac{e_i}{c} \mathbf{A}(\mathbf{x}_i)|^2 + V + \frac{1}{8\pi} \int d^3x(E^2 + B^2) \\ &= H_{\text{atom}} + H_{\text{rad}} + H_I \end{aligned} \tag{3-2}$$

where H_{atom} is given by Eq. 3-1, H_{rad} is given by Eqs. 2-12 and 2-19, and H_1 is the Hamiltonian for the interaction of the field and the atom. It is given by

$$H_I = \sum_i \left\{ -\frac{e_i}{m_i c} \mathbf{p}_i \cdot \mathbf{A}(\mathbf{x}_i) + \frac{e^2}{2m_i c^2} A^2(\mathbf{x}_i) \right\} \tag{3-3}$$

To simplify the notation we drop the subscript i and let H_I be the interaction Hamiltonian for only one of the particles with the field. The summation is easily reintroduced whenever it is needed.

We write

$$H_I = H' + H'' \tag{3-4}$$

where H' is the part proportional to A and H'' is the part proportional to A^2. Using Eq. 2.11 we find

$$H' = -\frac{e}{mc}\sum_{\mathbf{k},\sigma}\left(\frac{2\pi\hbar c^2}{\Omega\omega_k}\right)^{1/2}\mathbf{p}\cdot\mathbf{u}_{\mathbf{k}\sigma}[a_{\mathbf{k}\sigma}e^{i\mathbf{k}\cdot\mathbf{x}} + a^+_{\mathbf{k}\sigma}e^{-i\mathbf{k}\cdot\mathbf{x}}] \tag{3-5a}$$

$$\begin{aligned} H'' = {} & \frac{e^2}{2mc^2}\sum_{\mathbf{k},\sigma}\sum_{\mathbf{k}'\sigma'}\left(\frac{2\pi\hbar c^2}{\Omega}\right)\frac{\mathbf{u}_{\mathbf{k}\sigma}\cdot\mathbf{u}_{\mathbf{k}'\sigma'}}{(\omega_k\omega_k')^{1/2}} \\ & \times\{a_{\mathbf{k}\sigma}a_{\mathbf{k}'\sigma'}e^{i(\mathbf{k}+\mathbf{k}')\cdot\mathbf{x}} + a_{\mathbf{k}\sigma}a^+_{\mathbf{k}'\sigma'}e^{i(\mathbf{k}-\mathbf{k}')\cdot\mathbf{x}} \\ & + a^+_{\mathbf{k}\sigma}a_{\mathbf{k}'\sigma'}e^{i(-\mathbf{k}+\mathbf{k}')\cdot\mathbf{x}} + a^+_{\mathbf{k}\sigma}a^+_{\mathbf{k}'\sigma'}e^{i(-\mathbf{k}-\mathbf{k}')\cdot u}\} \end{aligned} \tag{3-5b}$$

The H_I will be treated as a perturbation. The unperturbed Hamiltonian, $H_0 = H_{\text{atom}} + H_{\text{rad}}$ has the eigenvectors

$$|\text{atom} + \text{radiation}\rangle = |a\rangle_{\text{atom}}\,|\cdots n_{\mathbf{k}\sigma}\cdots\rangle_{\text{rad}} \tag{3-6}$$

where we have let a stand for the quantum numbers of the atom. The H_I induces transitions between these states whose transition probability per unit time is given by Eq. 1-165.

It is clear from inspection of Eq. 3-8 that in first order perturbation theory H' induces transitions in which the number of photons changes by ± 1, since each term in H' contains one and only one creation or destruction operator. Similarly, H'' induces transitions in which two photons are emitted, two are absorbed or one is emitted and another is absorbed.

In the sections that follow we discuss some examples.

EMISSION OF LIGHT BY AN EXCITED ATOM

Consider an atom initially in state $|a\rangle_{\text{atom}}$ decaying to state $|b\rangle_{\text{atom}}$ with the emission of a photon with wave vector $\mathbf{k}$ and polarization σ. We write the initial and final states of H_0 as

$$|i\rangle = |a\rangle_{\text{atom}}\,|\cdots n_{\mathbf{k}\sigma}\cdots\rangle_{\text{rad}} \tag{3-7a}$$

$$|f\rangle = |b\rangle_{\text{atom}}\,|\cdots n_{\mathbf{k}\sigma} + 1\cdots\rangle_{\text{rad}} \tag{3-7b}$$

Only H' connects these states in the first order contribution to M_{fi}. We find

$$\langle f|\,H'\,|i\rangle = -\frac{e}{mc}\left(\frac{2\pi\hbar c^2}{\Omega\omega_k}\right)^{1/2}\langle b|\,\mathbf{p}\cdot\mathbf{u}_{\mathbf{k}\sigma}e^{i\mathbf{k}\cdot\mathbf{x}}\,|a\rangle_{\text{atom}}\cdot\sqrt{n_{\mathbf{k}\sigma}+1} \tag{3-8}$$

where Eq. 2-26a has been used. Note that of the terms in Eq. 3-5a none of the destruction operators and only one of the creation operators contribute to this matrix element. The energy difference between final and initial states is

$$E_f - E_i = E_b - E_a + \hbar\omega_k \tag{3-9}$$

According to Eq. 1-165 the transition probability per unit time given by first order perturbation theory is

$$\begin{aligned}\left(\frac{\text{trans prob}}{\text{time}}\right)_{\text{emiss}} &= \frac{2\pi}{\hbar}\,|\langle f|\,H'\,|i\rangle|^2\,\delta(E_f - E_i)\\ &= \frac{2\pi}{\hbar}\left(\frac{e}{mc}\right)^2\left(\frac{2\pi\hbar c^2}{\Omega\omega_k}\right)(n_{\mathbf{k}\sigma}+1)\,|\langle b|\,\mathbf{p}\cdot\mathbf{u}_{\mathbf{k}\sigma}e^{-i\mathbf{k}\cdot\mathbf{x}}\,|a\rangle|^2\\ &\quad\times\,\delta(E_b - E_a + \hbar\omega_k)\end{aligned} \tag{3-10}$$

Note the factor $n_{\mathbf{k}\sigma} + 1$. The term in Eq. 3-10 proportional to $n_{\mathbf{k}\sigma}$, the number of photons already present in the state into which the photon is emitted, is called stimulated emission. The term that remains when $n_{\mathbf{k}\sigma} = 0$ is called spontaneous emission. We consider spontaneous emission first. Stimulated emission may be treated together with absorption.

To calculate the lifetime of the excited state of an atom against spontaneous emission of a photon, we set $n_{\mathbf{k}\sigma} = 0$ and sum Eq. 3-10 over all of the $\mathbf{k}$'s and σ's that the emitted photon can have. That is,

$$\left(\frac{1}{\tau}\right)_{a\to b} = \frac{4\pi^2e^2}{m^2\Omega}\sum_{\mathbf{k},\sigma}\frac{1}{\omega_k}\,|\langle b|\,\mathbf{p}\cdot\mathbf{u}_{\mathbf{k}\sigma}e^{-i\mathbf{k}\cdot\mathbf{x}}\,|a\rangle|^2\cdot\delta(E_b - E_a + \hbar\omega_k) \tag{3-11}$$

Now we let the volume of the box in which the electromagnetic field is quantized become infinite. A very useful formula is

$$\sum_{\mathbf{k}} \xrightarrow[\Omega\to\infty]{} \frac{\Omega}{(2\pi)^3}\int d^3k \tag{3-12}$$

Problem 3-1. Prove Eq. 3-12. *Hint:* use $k_i = 2\pi n_i/L$ where n_i is an integer to show that the number of states with k_x in Δk_x, k_y in Δk_y and k_z in Δk_z is $L^3/(2\pi)^3\,\Delta k_x\,\Delta k_y\,\Delta k_z$. In the limit that $\Omega = L^3 \to \infty$ show that Eq. 3-12 results.

In doing the sum over polarizations we choose $\mathbf{u}_{\mathbf{k}1}$ and $u_{\mathbf{k}1}$ as shown in Fig. 3-1.

Then

$$\sum_{\sigma=1,2}|\langle b|\,\mathbf{p}\cdot\mathbf{u}_{\mathbf{k}\sigma}e^{-i\mathbf{k}\cdot\mathbf{x}}\,|a\rangle|^2 = |\langle b|\,\mathbf{p}e^{-i\mathbf{k}\cdot\mathbf{x}}\,|a\rangle|^2\sin^2\theta \tag{3-13}$$

For wavelengths of light which are much larger that atomic dimensions it is a good approximation to write

$$e^{-i\mathbf{k}\cdot\mathbf{x}} = 1 - \mathbf{k}\cdot\mathbf{x} + \tfrac{1}{2}(\mathbf{k}\cdot\mathbf{x})^2\,\ldots\ldots \tag{3-14}$$

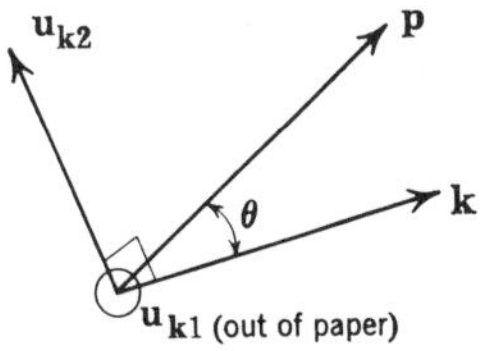

Figure 3-1

and keep only the first few terms. When only the first term is retained, it is called the electric dipole approximation for reasons that will soon become apparent. The higher terms give electric quadrupole, magnetic dipole, and so on. Making the electric dipole approximation and using Eqs. 3-12 and 3-13 give

$$\left(\frac{1}{\tau}\right)_{a\to b} = \frac{e^2}{2\pi m^2}\int d^3k\,\frac{1}{\omega_k}\,|\langle b|\,\mathbf{p}\,|a\rangle|^2 \sin^2\theta\cdot\delta(E_b - E_a + \hbar\omega_k) \quad (3\text{-}15)$$

Next we introduce spherical coordinates in k-space taking the k_z axis along the direction of $\langle b|\,\mathbf{p}\,|a\rangle$. Then

$$\begin{aligned} d^3k &= k^2\,dk\,\sin\theta\,d\theta\,d\Phi \\ &= 2\pi\,\frac{\omega_k^{\,2}\,d\omega_k}{c^3}\,\sin\theta\,d\theta \end{aligned} \quad (3\text{-}16)$$

Carrying out the integrations in Eq. 3-15 gives

$$\left(\frac{1}{\tau}\right)_{a\to b} = \frac{4e^2}{3m^2c^3\hbar}\,\omega_{ab}\,|\langle b|\,\mathbf{p}\,|a\rangle|^2 \quad (3\text{-}17)$$

where $\omega_{ab} = (E_a - E_b)/\hbar$ is the frequency of the emitted photon.

Equation 3-17 can be expressed in another form by using the Heisenberg equations of motion, Eq. 1-77, to write

$$\begin{aligned} \langle b|\,\mathbf{p}\,|a\rangle &= \langle b|\,m\,\frac{d\mathbf{x}}{dt}\,|a\rangle \\ &= -\frac{im}{\hbar}\,\langle b|\,\mathbf{x}H - H\mathbf{x}\,|a\rangle \\ &= \frac{im}{\hbar}\,(E_a - E_b)\langle b|\,\mathbf{x}\,|a\rangle = im\omega_{ab}\langle b|\,\mathbf{x}\,|a\rangle \end{aligned} \quad (3\text{-}18)$$

Equation 3-17 can then be written as

$$\left(\frac{1}{\tau}\right)_{a\to b} = \frac{4e^2\omega_{ab}^{\,3}}{3\hbar c^3}\,|\langle b|\,\mathbf{x}\,|a\rangle|^2 \quad (3\text{-}19)$$

In this form it is clear that the transition probability per unit time is proportional to the square of the matrix element of the dipole moment $e\mathbf{x}$ of the radiating electron.

Using the Heisenberg equations of motion to write

$$|\langle b|\frac{d^2\mathbf{x}}{dt^2}|a\rangle| = \omega_{ab}{}^4\, |\langle b|\,\mathbf{x}\,|a\rangle|^2 \tag{3-20}$$

Equation 3-19 can be written in the form

$$\left(\frac{\hbar\omega_{ab}}{\tau}\right)_{a\to b} = \frac{4e^2}{3c^3}|\langle b|\frac{d^2\mathbf{x}}{dt^2}|a\rangle|^2 \tag{3-21}$$

In this form our result bears a striking resemblance to the classical formula of Larmor which says that the energy radiated per unit time by an accelerated nonrelativistic electron is

$$\frac{\text{energy}}{\text{time}} = \frac{2e^2}{3c^3}\left|\frac{d^2\mathbf{x}}{dt^2}\right|^2 \tag{3-22}$$

Problem 3-2. Show that the selection rules for electric dipole transitions are $\Delta l = \pm 1$ and $\Delta m = \pm 1, 0$, where l and m are the angular momentum quantum numbers of the electron.

Problem 3-3. The $1s$, $2s$, and $2p$ wave functions of hydrogen are

$$\psi(1s) = \frac{1}{\sqrt{\pi a^3}}\, e^{-r/a} \tag{3-23a}$$

$$\psi(2s) = \frac{1}{4\sqrt{2\pi a^3}}(2 - r/a)e^{-r/a} \tag{3-23b}$$

$$\psi(2p) = \frac{1}{8\sqrt{\pi a^3}}\frac{r}{a}\, e^{-r/2a}\begin{cases}\sin\theta e^{i\phi} & m = 1\\ \sqrt{2}\cos\theta & m = 0\\ \sin\theta e^{-i\phi} & m = -1\end{cases} \tag{3-23c}$$

where $a = \hbar^2/me^2$ is the Bohr radius. Calculate τ for the $2p \to 1s$ transition.

Problem 3-4. Show that the $2s$ state of hydrogen cannot decay to the $1s$ state through the $\mathbf{p}\cdot\mathbf{A}$ interaction with the emission of one photon by showing that

$$\langle 2s|\,\mathbf{u}_{\mathbf{k}\sigma}\cdot\mathbf{p}e^{-i\mathbf{k}\cdot\mathbf{x}}\,|1s\rangle = 0$$

Problem 3-5. Because of its magnetic moment, an electron has an interaction with the electromagnetic field in addition to the interactions $H' \sim \mathbf{p}\cdot\mathbf{A}$ and $H'' \sim A^2$. The magnetic moment of the electron is $\boldsymbol{\mu} = (e\hbar/2mc)\boldsymbol{\sigma}$

where the components of $\boldsymbol{\sigma}$ are given in Eq. 1-85. Find the interaction Hamiltonian for this spin dependent part of the interaction.

Problem 3-6. The magnetic interaction between the spin of the electron and the spin of the nucleus gives rise to a splitting of the ground state level of the hydrogen atom. The photon emitted when a transition occurs between these states has a wavelength of 21 cm. It has never been observed in the laboratory but is well known to radio astronomers. Use the results of Problem 3-5 to calculate the lifetime for this transition.

Problem 3-7. Use the results of Problem 3-5 to calculate the lifetime of the 2*s* state of hydrogen assuming it decays to the ground state through the spin dependent interaction with the emission of one photon. As will be seen in Problem 3-10, the two photon decay process is much more rapid than this.

ABSORPTION OF LIGHT

We take the initial and final states to be

$$|i\rangle = |b\rangle_{\text{atom}} \, |\cdots n_{\mathbf{k}\sigma} \cdots\rangle_{\text{rad}} \tag{3-24a}$$

$$|f\rangle = |a\rangle_{\text{atom}} \, |\cdots n_{k} - 1 \cdots\rangle_{\text{rad}} \tag{3-24b}$$

Using Eq. 2-25a we find by a calculation similar to that of the last section that

$$\left(\frac{\text{trans prob}}{\text{time}}\right)_{\text{abs}} = \frac{2\pi}{\hbar}\left(\frac{e}{mc}\right)^2\left(\frac{2\pi\hbar c^2}{\Omega\omega_k}\right) n_{\mathbf{k}\sigma} \times |\langle a|\, \mathbf{p}\cdot\mathbf{u}_{\mathbf{k}\sigma}e^{+i\mathbf{k}\cdot\mathbf{x}}\,|b\rangle|^2 \; \delta(E_b + \hbar\omega_k - E_a) \tag{3-25}$$

Since by Eq. 1-27

$$\langle a|\, \mathbf{p}\cdot\mathbf{u}_{\mathbf{k}\sigma}e^{i\mathbf{k}\cdot\mathbf{x}}\,|b\rangle = \langle b|\, \mathbf{p}\cdot\mathbf{u}_{\mathbf{k}\sigma}e^{-i\mathbf{k}\cdot\mathbf{x}}\,|a\rangle^* \tag{3-26}$$

we see by comparing Eq. 3-25 with Eq. 3-10 that the transition probability per unit time for absorption is equal to that for stimulated emission.

The incident flux of photons of momentum $\hbar\mathbf{k}$ and polarization σ is

$$\text{flux} = \frac{n_{k\sigma}}{\Omega}\, c \tag{3-27}$$

Dividing Eq. 3-25 by the incident flux gives the cross section for the absorption of a photon of momentum $\hbar\mathbf{k}$ and polarization σ by an atom which makes a transition from $|b\rangle$ to $|a\rangle$; it is

$$\sigma_{b\to a}(\mathbf{k}, \sigma) = \frac{4\pi^2 e^2}{m^2\omega_k c}\, |\langle a|\, \mathbf{p}\cdot\mathbf{u}_{\mathbf{k}\sigma}e^{i\mathbf{k}\cdot\mathbf{x}}\,|b\rangle|^2 \cdot \delta(E_b + \hbar\omega_k - E_a) \tag{3-28}$$

This will be meaningful only if the incident radiation has a continuous spectrum so that Eq. 3-28 will be integrated over frequency to obtain the energy absorbed from an incident beam of radiation. Actually, of course, spectral lines are not infinitely sharp as is implied by the δ-function in $\sigma_{b\to a}$. The line is broadened by a variety of processes, one of which is discussed in a later section.

Problem 3-8. Consider the photoelectric emission of an electron from the ground state of a hydrogen atom. Assume that the incident photon is sufficiently energetic for the wave function of the ejected electron to be approximated by a plane wave. Assume that the photons momentum is along the x-axis, and its polarization vector is along the z-axis. Make the dipole approximation. Calculate the differential cross section for ejection of an electron into the element of solid angle $d\Omega$.

BLACK BODY SPECTRUM

Suppose that we have a collection of atoms in thermal equilibrium. Let N_b be the number in state $|b\rangle$ and N_a be the number in state $|a\rangle$. Transitions will occur between these states as the atoms emit and absorb photons from the radiation field. We can write

$$\frac{d}{dt} N_b = -N_b\left(\frac{\text{trans prob}}{\text{time}}\right)_{\text{abs}} + N_a\left(\frac{\text{trans prob}}{\text{time}}\right)_{\text{emiss}} \tag{3-29a}$$

$$\frac{d}{dt} N_a = -N_a\left(\frac{\text{trans prob}}{\text{time}}\right)_{\text{emiss}} + N_b\left(\frac{\text{trans prob}}{\text{time}}\right)_{\text{abs}} \tag{3-29b}$$

In equilibrium we must have

$$\frac{d}{dt} N_b = \frac{d}{dt} N_a = 0 \tag{3-30}$$

and

$$\frac{N_b}{N_a} = e^{-(E_b-E_a)/kT} \tag{3-31}$$

It follows that

$$\frac{N_b}{N_a} = e^{\hbar\omega_k/kT} = \frac{(\text{trans prob/time})_{\text{emiss}}}{(\text{trans prob/time})_{\text{abs}}} = \frac{n_{\mathbf{k}\sigma}+1}{n_{\mathbf{k}\sigma}} \tag{3-32}$$

Solving for $n_{\mathbf{k}\sigma}$ gives

$$n_{\mathbf{k}\sigma} = \frac{1}{e^{\hbar\omega_k/kT} - 1} \tag{3-33}$$

which is the Planck distribution. From this we may obtain the energy per unit volume with $\mathbf{k}$ in d^3k as

$$2\frac{\hbar\omega_k n_{k\sigma}}{N}\cdot\Omega\frac{d^3k}{(2\pi)^3} \tag{3-34}$$

If we write $u(\omega)\,d\omega$ as the energy per unit volume with ω in $d\omega$, then from Eq. 3-33 we find

$$u(\omega) = \frac{\hbar\omega^3}{8c^3}\frac{1}{e^{\hbar\omega/kT} - 1} \tag{3-35}$$

SCATTERING OF LIGHT BY A FREE ELECTRON

If we let $V = 0$ in Eq. 3-1, then the Hamiltonian $p^2/2m$ for a free electron has the eigenvector $|\mathbf{q}\rangle$ where

$$\langle \mathbf{x} \mid \mathbf{q}\rangle = \psi_{\mathbf{q}}(\mathbf{x}) = \frac{1}{\sqrt{\Omega}}\,e^{i\mathbf{q}\cdot\mathbf{x}} \tag{3-36}$$

These are normalized so that

$$\int_{\Omega} d^3x\,\psi_{\mathbf{q}}^*(x)\psi_{\mathbf{q}'}(\mathbf{x}) = \delta_{\mathbf{q},\mathbf{q}'} \tag{3-37}$$

The energy eigenvalues are $E_{\mathbf{q}} = \hbar^2q^2/2m$.

It is easily shown that in free space a free electron cannot emit or absorb a photon without violating energy or momentum conservation. Therefore there are no first-order processes involving H'. However, there are first-order processes involving H''. We shall consider the scattering of light. In this process one photon is destroyed and another is created. It can be pictured schematically by the Feynmann diagram of Fig. 3-2.

At the vertex an electron changes its state from $|\mathbf{q}_i\rangle$ to $|\mathbf{q}_f\rangle$ and a photon with momentum $h\mathbf{k}_i$ and polarization σ_i is destroyed and another with

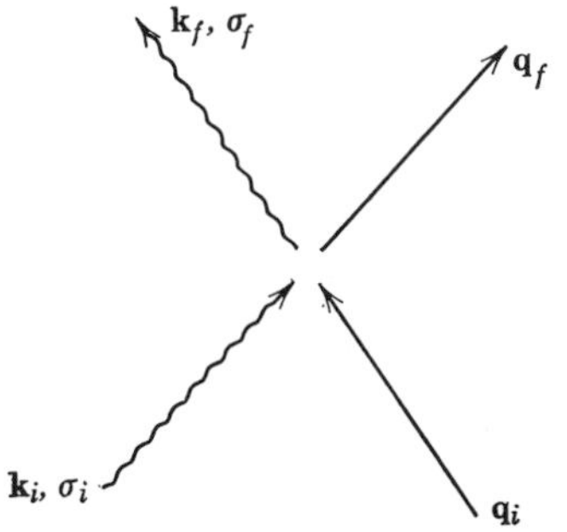

Figure 3-2

momentum $\hbar\mathbf{k}_f$ and polarization σ_f is created. The initial and final states are

$$|i\rangle = |\mathbf{q}_i\rangle_e \, |\cdots n_{\mathbf{k}_i\sigma_i} \cdots n_{\mathbf{k}_f\sigma_f} \cdots\rangle_{\text{rad}} \tag{3-38a}$$

$$|f\rangle = |\mathbf{q}_f\rangle_e \, |\cdots n_{\mathbf{k}_i\sigma_i} - 1, \cdots n_{\mathbf{k}_f\sigma_f} + 1, \cdots\rangle_{\text{rad}} \tag{3-38b}$$

This transition from $|i\rangle$ to $|f\rangle$ can be produced by the term in Eq. 3-5b containing $a_{\mathbf{k}_i\sigma_i} a^+_{\mathbf{k}_f\sigma_f}$. The transition probability per unit time is

$$\left(\frac{\text{trans prob}}{\text{time}}\right)_{\text{scatt}} = \frac{2\pi}{\hbar}\left(\frac{e^2}{2mc^2}\right)^2\left(\frac{2\pi\hbar c^2}{\Omega}\right)^2 \times 2^2 \frac{|\mathbf{u}_i \cdot \mathbf{u}_f|^2}{\omega_i\omega_f} |\langle\mathbf{q}_f|\, e^{i(\mathbf{k}_i - \mathbf{k}_f)\cdot\mathbf{x}}\, |\mathbf{q}_i\rangle|^2 \times n_i(n_f + 1)\, \delta\left[\hbar\omega_i + \frac{\hbar^2 q_i^{\,2}}{2m} - \hbar\omega_f - \frac{\hbar^2 q_f^{\,2}}{2m}\right] \tag{3-39}$$

We have simplified the notation somewhat by replacing the subscripts $\mathbf{k}_i$, σ_i by i and $\mathbf{k}_f\sigma_f$ by f. The factors n_i and $(n_f + 1)$ come from the matrix elements of a_i and a_f^+ when Eqs. 2-25 are used. The factor 2^2 comes from using both the second and third terms in Eq. 3-5b. The matrix element in Eq. 3-39 is

$$\langle\mathbf{q}_f|\, e^{i(\mathbf{k}_i - \mathbf{k}_f)\cdot\mathbf{x}}\, |\mathbf{q}_i\rangle = \int \frac{d^3x}{\Omega}\, e^{i(\mathbf{q}_i + \mathbf{k}_i - \mathbf{q}_f - \mathbf{k}_f)\cdot\mathbf{x}} = \delta_{\mathbf{q}_i + \mathbf{k}_i, \mathbf{q}_f + \mathbf{k}_f} \tag{3-40}$$

This shows that the transition probability vanishes unless the initial momentum $\hbar(\mathbf{q}_i + \mathbf{k}_i)$ is equal to the final momentum $\hbar(\mathbf{q}_f + \mathbf{k}_f)$. The term in n_f shows that it is possible to have "stimulated" scattering; that is, the scattering is enhanced if there are photons present in the final state. For our present purposes we assume that this is not the case and set $n_f = 0$.

The Kronecker-δ in Eq. 3-39 show that both momentum and energy are conserved in the scattering process. This is enough to derive the frequency shift of the scattered photon. To do it properly one should replace the nonrelativistic energies $\hbar^2q^2/2m$ by the relativistic energies $\sqrt{\hbar^2q^2c^2 + m^2c^4}$.

Problem 3-9. Use the nonrelativistic conservation laws to show that when the shift in wavelength is small it is given by

$$\lambda_f - \lambda_i = \Delta\lambda = (h/mc)(1 - \cos\theta) \tag{3-41}$$

where $\cos\theta = \mathbf{k}_i \cdot \mathbf{k}_f/k_ik_f$. Assume that the initial velocity of the electron is zero.

The total scattering cross section can be obtained by summing Eq. 3-39 over the final states of both photon and electron and equating the result to the product of the cross section and the incident flux given by Eq. 3-27. Thus

$$\sigma_T\left(\frac{n_i c}{\Omega}\right) = \frac{2\pi}{\hbar}\left(\frac{e^2}{2mc^2}\right)^2\left(\frac{2\pi\hbar c^2}{\Omega}\right)^2 \cdot 4\sum_{\mathbf{k}_f\sigma_f}\sum_{\mathbf{q}_f}\frac{|\mathbf{u}_i\cdot\mathbf{u}_f|^2}{\omega_i\omega_f}\, n_i\,\delta_{\mathbf{q}_i+\mathbf{k}_i,\mathbf{q}_f+\mathbf{k}_f}$$
$$\times\,\delta\left[\hbar\omega_i + \frac{\hbar^2 q_i^{\,2}}{2m} - \hbar\omega_f - \frac{\hbar^2 q_f^{\,2}}{2m}\right] \qquad (3\text{-}42)$$

Using the Kronecker-δ to get rid of the sum over $\mathbf{q}_f$ and using Eq. 3-12 we obtain

$$\sigma_T = \frac{e^4\hbar}{m^2c}\sum_{\sigma_f}\int d^3k_f\,\frac{|\mathbf{u}_i\cdot\mathbf{u}_f|^2}{\omega_i\omega_f}$$
$$\times\,\delta\left[\hbar c(k_i - k_f) + \frac{\hbar^2}{2m}q_i^{\,2} - \frac{\hbar^2}{2m}|\mathbf{q}_i + \mathbf{k}_i - \mathbf{k}_f|^2\right] \qquad (3\text{-}43)$$

As may be seen from Eq. 3-41 the wavelength shift is a quantum-mechanical effect. If it is neglected, then the δ-function in Eq. 3-43 is approximately

$$\delta[\hbar c(k_i - k_f)] = \frac{1}{\hbar c}\,\delta(k_i - k_f) \qquad (3\text{-}44)$$

and σ_T becomes

$$\sigma_T = \left(\frac{e^2}{mc^2}\right)^2\sum_{\sigma_f}\int d\Omega_f\,|\mathbf{u}_i\cdot\mathbf{u}_f|^2 \qquad (3\text{-}45)$$

where $d\Omega_f$ is the element of solid angle into which the photon is scattered. We may interpret

$$\frac{d\sigma_T}{d\Omega_f} = \left(\frac{e^2}{mc^2}\right)^2|\mathbf{u}_i\cdot\mathbf{u}_f|^2 \qquad (3\text{-}46)$$

as the differential cross section for scattering a photon polarized in the direction $\mathbf{u}_i$ into $d\Omega_f$ with polarization $\mathbf{u}_f$.

We may get the cross section for scattering of unpolarized light by averaging Eq. 3-45 over initial polarizations. Using

$$\tfrac{1}{2}\sum_{\sigma_i}\sum_{\sigma_f}|\mathbf{u}_i\cdot\mathbf{u}_f|^2 = \tfrac{1}{2}(1 + \cos^2\theta) \qquad (3\text{-}47)$$

we carry out the integration over angles and obtain

$$\sigma_T = \frac{8\pi}{3}\left(\frac{e^2}{mc^2}\right)^2 = \frac{8\pi}{3}r_e^{\,2} \qquad (3\text{-}48)$$

This is the Thompson cross section which may be obtained classically. The quantity $r_e = e^2/mc^2$ is the classical radius of the electron.

Problem 3-10. Calculate the lifetime of the $2s$ state of hydrogen assuming that it decays by two-photon emission. Note that it is necessary to combine the second order matrix element of $H' \sim \mathbf{p} \cdot \mathbf{A}$ with the first order matrix element of $H'' \sim A^2$. Do not try to do the problem exactly but obtain an order of magnitude estimate of the lifetime.

ČERENKOV RADIATION

As we have stated previously, a free electron moving in a vacuum cannot emit or absorb a photon and still conserve momentum and energy. However, a particle moving through a dielectric medium can have a velocity greater than the velocity of light in this medium, and under these circumstances, as we shall show, it is possible for it to emit or absorb a photon. A dielectric is characterized by its dielectric constant $\varepsilon(\omega)$, and its index of refraction is given by $n(\omega) = \sqrt{\varepsilon(\omega)}$. We assume that these are frequency-dependent quantities. The relation between frequency and wave number is

$$\omega = \frac{c}{n(\omega)} k = \frac{c}{\sqrt{\varepsilon(\omega)}} k \tag{3-49}$$

The calculation of the energy of the electromagnetic field given in Chapter 2 is still valid but this energy is not the total energy associated with the wave. The particles of the medium move in response to the wave and their energy must be properly included in the total energy. Landau and Lifschitz[15] have shown that in such a dielectric medium the energy is

$$U = \int d^3x \frac{1}{8\pi}\left\{|\mathbf{E}|^2 \frac{\partial}{\partial\omega} \omega\varepsilon(\omega) + |\mathbf{B}|^2\right\} \tag{3-50}$$

Since

$$\nabla \times \mathbf{E} = -\frac{1}{c}\frac{\partial \mathbf{B}}{\partial t}$$

$$i\mathbf{k} \times \mathbf{E} = \frac{i\omega}{c} \mathbf{B} \tag{3-51}$$

$$|\mathbf{B}|^2 = \frac{c^2}{\omega^2} |\mathbf{k} \times \mathbf{E}|^2 = \frac{c^2k^2}{\omega^2} |\mathbf{E}|^2 = \varepsilon\, |E|^2$$

Eq. 3-50 becomes

$$U = \int d^3x \frac{1}{8\pi} |\mathbf{E}|^2 \left[\frac{\partial}{\partial\omega} \omega\varepsilon + \varepsilon\right]$$

$$= \int d^3x \frac{1}{4\pi} |\mathbf{E}|^2 \frac{1}{2\omega} \frac{\partial}{\partial\omega} \omega^2\varepsilon \tag{3-52}$$

We see that the energy density that a wave would have in a vacuum must be corrected by the factor

$$\frac{1}{2\omega}\frac{\partial}{\partial\omega}\,\omega^2\varepsilon(\omega) \tag{3-53}$$

when it moves in a medium of dielectric constant $\varepsilon(\omega)$. We want the total energy rather than just the electromagnetic field energy to have the form of Eq. 2-19, so that it can be interpreted as the sum of the energies of harmonic oscillators. In order to accomplish this we must modify the normalization factor in Eq. 2-6 so that when the energy of each oscillator is corrected by the factor of Eq. 3-62 it becomes $\hbar\omega_k a_{\mathbf{k}\sigma}^+ a_{\mathbf{k}\sigma}$. This is accomplished if we choose the normalization factor in Eq. 2-6 to be

$$\left\{\frac{2\pi\hbar c^2}{\Omega[(1/2\omega)(\partial/\partial\omega)(\omega^2\varepsilon)]_{\omega_k}}\right\}^{1/2} \tag{3-54}$$

Equation 2-11a becomes

$$\mathbf{A}(\mathbf{x}, t) = \sum_{\mathbf{k},\sigma}\left\{\frac{2\pi\hbar c^2}{\Omega[(1/2\omega)(\partial/\partial\omega)(\omega^2\varepsilon)]_{\omega_k}}\right\}^{1/2}[a_{\mathbf{k}\sigma}e^{i\mathbf{k}\cdot\mathbf{x}} + a_{k\sigma}^+e^{-i\mathbf{k}\cdot\mathbf{x}}] \tag{3-55}$$

The interaction Hamiltonian H' in Eq. 3-5a is unchanged except for the change in the normalization factor.

Now, we calculate the transition probability per unit time for a free electron of momentum $\hbar\mathbf{q}$ to emit a photon of momentum $\hbar\mathbf{k}$ thereby changing its momentum to $\hbar(\mathbf{q} - \mathbf{k})$. We find

$$\begin{aligned}\left(\frac{\text{trans prob}}{\text{time}}\right)_{\mathbf{q}\to\mathbf{q}-\mathbf{k}} &= \frac{2\pi}{\hbar}\left(\frac{e}{mc}\right)^2\left[\frac{2\pi\hbar c^2}{\Omega[(1/2\omega)(\partial/\partial\omega)\omega\varepsilon]_{\omega_k}}\right] \\ &\times |\langle\mathbf{q} - \mathbf{k}|\,\mathbf{p}\cdot\mathbf{u}_{k\sigma}e^{-i\mathbf{k}\cdot\mathbf{x}}\,|\mathbf{q}\rangle|^2 \\ &\times \delta\left[\frac{\hbar^2q^2}{2m} - \frac{\hbar^2}{2m}|\mathbf{q} - \mathbf{k}|^2 - \hbar\omega_k\right]\end{aligned} \tag{3-56}$$

The matrix element in Eq. 3-56 is just equal to $\hbar\mathbf{q}\cdot u_{\mathbf{k}\sigma}$. Letting θ be the angle between $\mathbf{q}$ and $\mathbf{k}$ and letting $\mathbf{v} = \hbar\mathbf{q}/m$ be the particles velocity we find

$$\left(\frac{\text{trans prob}}{\text{time}}\right)_{\mathbf{q}\to\mathbf{q}-\mathbf{k}} = \frac{4\pi^2e^2\hbar^2\,|\mathbf{q}\cdot\mathbf{u}_{\ \sigma}|^2}{m^2\Omega\hbar vk[(1/2\omega)(\partial/\partial\omega)\omega\varepsilon]_{\omega_k}}\,\delta\left[\cos\theta - \frac{c}{nv} - \frac{\hbar\omega n}{2mcv}\right] \tag{3-57}$$

Note that the photon is emitted at an angle to the path of the electron given by

$$\cos\theta = \frac{c}{nv}\left[1 + \frac{\hbar\omega n^2}{2mc^2}\right] \tag{3-58}$$

If the energy of the photon $\hbar\omega$ is much less than the rest mass of the electron mc^2 then this is approximately $\cos\theta = c/nv$ which gives the classical Čerenkov angle. This can only be satisfied if the velocity of the particle is greater than c/n the velocity of the wave. In a vacuum where $n = 1$, v can never exceed c and so emission cannot occur.

A quantity of physical interest is the loss of energy per unit length of path of the electron. It is given by

$$\frac{dW}{dx} = \frac{1}{v}\frac{dW}{dt} = \frac{1}{v}\sum_{\mathbf{k},\sigma} \hbar\omega_k \left(\frac{\text{trans prob}}{\text{time}}\right)_{\mathbf{q}\to\mathbf{q}-\mathbf{k}} \tag{3-59}$$

Using

$$\sum_{\sigma} |\mathbf{q}\cdot\mathbf{u}_{\mathbf{k}\sigma}|^2 = q^2(1-\cos^2\theta) = \frac{m^2v^2}{\hbar^2}(1-\cos^2\theta) \tag{3-60}$$

and Eq. 3-12 and introducing spherical coordinates in k-space we find

$$\begin{aligned}\frac{dW}{dx} &= e^2\int_0^\infty k\,dk\int_{-1}^{+1} d(\cos\theta)\frac{(1-\cos^2\theta)\,\delta[\cos\theta-(c/nv)-(\hbar\omega n/2mcv)]}{[(1/2\omega)(\partial/\partial\omega)\omega^2\varepsilon]_{nkc}} \\ &= \frac{e^2}{c^2}\int\frac{\varepsilon(\omega)\omega^2\,d\omega}{[\frac{1}{2}(\partial/\partial\omega)\omega^2\varepsilon]}\left[1-\frac{c^2}{n^2v^2}\left(1+\frac{\hbar\omega n^2}{2mc^2}\right)^2\right]\end{aligned} \tag{3-61}$$

It is clear from this derivation that the integration over ω is only over those frequencies for which Eq. 3-58 can be satisfied. Since

$$n(\omega) \xrightarrow[\omega\to\infty]{} 1 \tag{3-62}$$

the range of integration does not extend to infinity and the integral is convergent.

Problem 3-11. Show that if relativistic expressions for the particle energies are used rather than nonrelativistic expressions, the Čerenkov angle is given by

$$\cos\theta = \frac{c}{nv}\left[1+\frac{\hbar\omega}{2mc^2}(n^2-1)\sqrt{1-v^2/c^2}\right] \tag{3-63}$$

rather than Eq. 3-58.

NATURAL LINE WIDTH

When an atom emits light, the emitted wave train is of finite duration. When this wave train of finite length is Fourier analyzed one finds a spectral line of finite width. However, when we calculated the emission and absorption of light earlier in this chapter we found infinitely sharp spectral lines. This is indicated by the presence of $\delta(E_a - E_b - \hbar\omega_k)$ in formulas such as Eqs. 3-10 and 3-28. The fault may be traced back to the assumptions made in the perturbation theory of Chapter 1. A simple modification in the perturbation theory will correct this and lead to a finite line width.[16]

We reconsider the emission of light by an atom. We assume that the atom is initially in state $|a\rangle$, and, for simplicity we assume that there is only one other state $|b\rangle$ into which it can decay. We assume no photons in the initial state and one photon in the final state. The initial and final states are then

$$|i\rangle = |a\rangle\, |\text{no photons}\rangle \tag{3-64a}$$

$$|f\rangle = |b\rangle\, |\cdots 1_{\mathbf{k}\sigma} \cdots\rangle \tag{3-64b}$$

In Eq. 1-158 we derived differential equations for the amplitudes of states. We denote the amplitudes of $|i\rangle$ and $|f\rangle$ C_{ao} an $C_{b\mathbf{k}\sigma}$. In deriving Eq. 1-160 we assumed that the amplitude of the initial state did not depart appreciably from unity. It is this assumption which must be modified. We write the differential equations for the amplitudes as

$$\frac{d}{dt} C_{ao} = -\frac{i}{\hbar} \sum_{\mathbf{k},\sigma} \langle a, o|\, H'\, |b, \mathbf{k}\sigma\rangle e^{i/\hbar(E_a - E_b - \ \omega_k)t} C_{b\mathbf{k}\sigma} \tag{3-65a}$$

$$\frac{d}{dt} C_{b\mathbf{k}\sigma} = \frac{i}{h} \langle b, \mathbf{k}, \sigma|\, H'\, |a, o\rangle e^{-i/\hbar(E_a - E_b - \hbar\omega_k)t} C_{ao} \tag{3-65b}$$

In Eq. 3-74a we retained on the right-hand side all of the states of the radiation field into which the initial state could decay. In Eq. 3-65b we retained only the term proportional to C_{ao} on the assumption that all other amplitudes remained negligibly small. Now let us assume that the decay of the initial state is exponential so

$$C_{ao} = e^{-\gamma/2t} \tag{3-66}$$

where γ is still to be determined. Using this in Eq. 3-65b and integrating from time zero to time t gives

$$C_{b\mathbf{k}\sigma}(t) = \langle b, \mathbf{k}\sigma|\, H'\, |a, o\rangle \frac{e^{-i/\hbar[(E_a - E_b - \ \omega_k) - i\gamma/2]t} - 1}{[(E_a - E_b - \hbar\omega_k) - i\gamma/2]} \tag{3-67}$$

After a lapse of a long period of time (more precisely $\gamma t \gg 1$), we find that

$$|C_{b\mathbf{k}\sigma}|^2 = \frac{1}{\hbar^2} |\langle b, \mathbf{k}\sigma|\, H'\, |a, o\rangle|^2 \frac{1}{(\omega_k - \omega_{ab})^2 + \gamma^2/4} \tag{3-68}$$

This is the probability of finding a photon of frequency ω_k in the radiation field, hence it gives the intensity distribution in the emitted line. The line is seen to have a Lorentz shape centered on $\omega_{ab} = (E_a - E_b)/\hbar$ with a half-width of $\gamma/2$.

We must choose γ so that Eq. 3-65a is satisfied. Substituting Eqs. 3-66 and 3-67 into 3-65a gives

$$\frac{-i\hbar\gamma}{2} = \sum_{\mathbf{k},\sigma} |\langle b, \mathbf{k}\sigma| H' |a, o\rangle|^2 \left[\frac{1 - e^{+i/\hbar[(E_a - E_b - \hbar\omega_k) - i\gamma/2]t}}{(E_a - E_b - \hbar\omega_k) - i\gamma/2}\right] \qquad (3\text{-}69)$$

If we neglect γ on the right-hand side of Eq. 3-78 we may write

$$\frac{1 - e^{i(\omega_{ab} - \omega_k)t}}{(\omega_{ab} - \omega_k)} = \frac{1 - \cos(\omega_{ab} - \omega_k)t}{(\omega_{ab} - \omega_k)} - i\,\frac{\sin(\omega_{ab} - \omega_k)t}{(\omega_{ab} - \omega_k)} \qquad (3\text{-}70)$$

Now, for large times $\sin \omega t/\omega$ is a function of ω which is very sharply peaked about $\omega = 0$; the area under the curve is π, so we may say that

$$\frac{\sin \omega t}{\omega} \xrightarrow[t\to\infty]{} \pi\,\delta(\omega_{ab} - \omega_k) \qquad (3\text{-}71)$$

Using this in Eq. 3-78 gives the real part of γ as

$$\operatorname{Re} \gamma = \frac{2\pi}{\hbar} \sum_{\mathbf{k},\sigma} |\langle b, \mathbf{k}\sigma| H' |a, o\rangle|^2\, \delta(E_a - E_b - h\omega_k) \qquad (3\text{-}72)$$

This is just the total transition probability per unit time. We previously called it $1/\tau$ where τ is the lifetime of the state $|a\rangle$. Hence

$$\gamma = \frac{1}{\tau} \qquad (3\text{-}73)$$

which is what we expect.

There is also an imaginary part of γ which comes from the real part of Eq. 3-70. This implies a shift in the frequency of the spectral line due to an interaction with the radiation field. Indeed, there is such a shift. We discuss it in the last chapter where a more careful treatment can be given.

4

Second Quantization

In the preceding chapters we have seen how the classical radiation field assumed characteristics describable in particle language when the electromagnetic field was quantized. This suggests the possibility that all of the particles found in nature may be considered as the quanta of some field. But what field? A natural choice is the wave function $\psi(\mathbf{x}, t)$ which describes the particle. We begin with the nonrelativistic Schrödinger equation

$$-\frac{\hbar}{i}\frac{\partial\psi}{\partial t} = -\frac{\hbar^2}{2m}\nabla^2\psi + V(\mathbf{x})\psi \tag{4-1}$$

for a particle in the presence of a potential $V(\mathbf{x})$. This is the equation that we quantize in this chapter. In a later chapter we discuss the quantization of the Dirac equation.

Let $\psi_n(\mathbf{x})$ be the solution of

$$\left(-\frac{\hbar^2}{2m}\nabla^2 + V\right)\psi_n = E_n\psi_n \tag{4-2}$$

and write

$$\psi(\mathbf{x}, t) = \sum_n b_n(t)\psi_n(\mathbf{x}) \tag{4-3}$$

From Eq. 4-1 we find

$$\frac{d}{dt}b_n = -\frac{i}{\hbar}E_n b_n \tag{4-4}$$

We would like to find a Hamiltonian that yields Eq. 4-4 as the equation of motion. A natural guess is

$$H = \int d^3x\psi^*(\mathbf{x}, t)\left[-\frac{\hbar^2}{2m}\nabla^2 + V\right]\psi(\mathbf{x}, t) \tag{4-5}$$

since this is the expectation value of the energy. Substituting Eq. 4-3 and using Eq. 4-2 and the orthonormality of the ψ_n's we obtain

$$H = \sum_n E_n b_n^* b_n \tag{4-6}$$

This looks like the Hamiltonian for a collection of harmonic oscillators with frequencies $E_n/\hbar$. If we interpret b_n as an operator, and b_n^*, which we now call b_n^+, as its adjoint and assume the commutation relations

$$[b_n, b_{n'}]_- = [b_n^+, b_{n'}^+]_- = 0 \tag{4-7a}$$

$$[b_n, b_{n'}^+]_- = \delta_{nn'} \tag{4-7b}$$

then the Heisenberg equations of motion

$$-\frac{\hbar}{i}\frac{d}{dt} b_n = [b_n, H]_- \tag{4-8}$$

give Eq. 4-4. Just as in Chapter 2, we arrive at a theory of quanta of the field that obey Bose-Einstein statistics. The operator $b_n^+ b_n$ has the eigenvalues $N_n = 0, 1, 2, 3, \ldots, \infty$, indicating that any integral number of particles may occupy the state whose wave function is ψ_n. The eigenvalues of H are

$$E = \sum_n E_n N_n \tag{4-9}$$

This is not entirely satisfactory, of course, since some of the particles found in nature obey Fermi-Dirac rather than Bose-Einstein statistics. We must look for some way of modifying the formalism so as to obtain a theory which describes Fermions. We wish to keep

$$H = \sum_n E_n b_n^+ b_n \tag{4-10}$$

as the Hamiltonian, and we want the Heisenberg equations of motion, Eq. 4-8, to yield Eq. 4-4. A little experimentation shows that Eq. 4-7 is not the only choice leading from Eq. 4-8 to Eq. 4-4. We could also assume

$$[b_n, b_{n'}]_+ = [b_n^+, b_{n'}^+]_+ = 0 \tag{4-11a}$$

$$[b_n, b_{n'}^+]_+ = \delta_{n,n'} \tag{4-11b}$$

where the anticommutator brackets are defined by

$$[A, B]_+ = AB + BA \tag{4-12}$$

(in anticipation of the introduction of anticommutator brackets we have put a minus as subscript on the commutator brackets in Eqs. 4-7 and 4-8.)

Using Eqs. 4-11 in Eq. 4-8 gives

$$\begin{aligned}
-\frac{\hbar}{i}\frac{d}{dt}b_n &= \sum_m E_m\{b_n b_m^+ b_m - b_m^+ b_m b_n\} \\
&= \sum_m E_m\{\delta_{nm} b_m - b_m^+ b_n b_m - b_m^+ b_m b_n\} \\
&= \sum_m E_m\{\delta_{nm} b_m + b_m^+ b_m b_n - b_m^+ b_m b_n\} \\
&= E_n b_n
\end{aligned} \tag{4-13}$$

In agreement with Eq. 4-4.

Let us now find the eigenvalues of $b_n^+ b_n$. Note that

$$\begin{aligned}
(b_n^+ b_n)(b_n^+ b_n) &= b_n^+(1 - b_n^+ b_n) b_n \\
&= b_n^+ b_n - b_n^+ b_n^+ b_n b_n \\
&= b_n^+ b_n
\end{aligned} \tag{4-14}$$

If λ is an eigenvalue of $b_n^+ b_n$, then

$$b_n^+ b_n |\lambda\rangle = \lambda |\lambda\rangle \tag{4-15a}$$

$$b_n^+ b_n b_n^+ b_n |\lambda\rangle = \lambda^2 |\lambda\rangle = \lambda |\lambda\rangle \tag{4-15b}$$

so $\lambda^2 = \lambda$. This is satisfied only for $\lambda = 0$ and $\lambda = 1$; thus these are the eigenvalues of the number operator $b_n^+ b_n$. We see that at most one particle can occupy the state $|n\rangle$. These particles obey Fermi-Dirac statistics.

We need the matrix elements of b_n and b_n^+. Write

$$b_n^+ b_n |N_n\rangle = N_n |N_n\rangle \tag{4-16}$$

where $N_n = 0, 1$. Consider the vector $b_n^+ |N_n\rangle$. Operating with $b_n^+ b_n$ gives

$$b_n^+ b_n b_n^+ |N_n\rangle = b_n^+(1 - b_n^+ b_n) |N_n\rangle = (1 - N_n) b_n^+ |N_n\rangle \tag{4-17}$$

We see that $b_n^+ |N_n\rangle$ is an eigenvector of $b_n^+ b_n$ with the eigenvalue $1 - N_n$, so we can write

$$b_n^+ |N_n\rangle = C_n |1 - N_n\rangle \tag{4-18}$$

The normalization constant can be found by taking the scalar product with $\langle N_n| b_n$ to obtain

$$\langle N_n| b_n b_n^+ |N_n\rangle = \langle N_n| 1 - b_n^+ b_n |N_n\rangle = (1 - N_n) = C_n^* C_n \tag{4-19}$$

from which

$$C_n = \theta_n \sqrt{1 - N_n} \tag{4-20}$$

where θ_n is a phase factor of modulus unity. A similar calculation gives

$$b_n |N_n\rangle = \theta_n \sqrt{N_n}\, |1 - N_n\rangle \tag{4-21}$$

where the phase factor can be chosen to be the same as in Eq. 4-20. The Fermion states can be written as

$$|\cdots N_n \cdots N_{n'} \cdots\rangle = |\cdots\rangle \cdots |N_n\rangle \cdots |N_{n'}\rangle \cdots \tag{4-22}$$

just as we did for the Boson states in Eq. 2-23.

It is convenient at this point to summarize and compare the following relations for Bosons and Fermions:

1. For Bosons:

$$b_n |\cdots N_n \cdots\rangle = \sqrt{N_n}\, |\cdots, N_n - 1, \cdots\rangle \tag{4-23a}$$

$$b_n^+ |\cdots N_n \cdots\rangle = \sqrt{N_n + 1}\, |\cdots, N_n + 1, \cdots\rangle \tag{4-23b}$$

2. For Fermions:

$$b_n |\cdots N_n \cdots\rangle = \theta_n \sqrt{N_n}\, |\cdots, 1 - N_n, \cdots\rangle \tag{4-24a}$$

$$b_n^+ |\cdots N_n \cdots\rangle = \theta_n \sqrt{1 - N_n}\, |\cdots, 1 - N_n, \cdots\rangle \tag{4-24b}$$

In both cases b_n is a destruction operator since it decreases the number of particles by one; b_n^+ is a creation operator in both cases. In the case of Bosons it was possible to choose the phase factors to be unity. The case of Fermions is somewhat more complicated. We want to choose the θ_n's so that

$$b_n b_{n'} + b_{n'} b_n$$

and

$$b_n^+ b_{n'}^+ + b_{n'}^+ b_n^+$$

give zero when these operators operate on any state vector. It may be shown that this is accomplished if the N_n's in Eq. 4-22 are ordered in some arbitrary way and then

$$\theta_n = (-1)^{\sum_{i=1}^{n-1}(N_i)} \tag{4-25}$$

For almost all applications in this book the θ_n's will play no role since only $|\theta_n|^2 = 1$ will enter the formulas.

There is a very general theorem due to Pauli that particles of integral spin obey Bose-Einstein statistics and particles of half-integral spin obey Fermi-Dirac statistics. The quantum numbers n in the wave function ψ_n for Fermions must be assumed to include the spin quantum numbers.

THE CONNECTION WITH ELEMENTARY QUANTUM MECHANICS

It might be thought that the process of second quantization applied to a quantum-mechanical equation such as Eq. 4-1 would endow the particles with properties of a quantum-mechanical nature which they did not previously possess. This is not the case. As we show in this section the elementary theory is contained in the second quantized theory. However, the second quantized theory possesses a flexibility that allows it to be extended to processes such as β-decay in which particles are destroyed and created.

According to Eq. 4-3, $\psi(\mathbf{x}, t)$ is a linear combination of the destruction operators b_n. We may interpret it as an operator which destroys a particle at the position $\mathbf{x}$ at the time t. Similarly, $\psi^+(\mathbf{x}, t)$ is a linear combination of the creation operators b_n^+. We may interpret it as an operator which creates a particle at the position $\mathbf{x}$ at the time t. The commutation relations for ψ and ψ^+ may be found from those for b_n and b_n^+. Thus

$$\begin{aligned}[\psi(\mathbf{x}, t), \psi^+(\mathbf{x}', t)]_\pm &= \sum_n \sum_{n'} [b_n, b_{n'}^+]_\pm \psi_n(\mathbf{x})\psi_n^*(\mathbf{x}') \\ &= \sum_n \psi_n(\mathbf{x})\psi_{n'}(\mathbf{x}') = \delta(\mathbf{x} - \mathbf{x}') \qquad (4\text{-}26)\end{aligned}$$

where Eqs. 4-7 and 4-11 and the completeness relation for the set of functions ψ_n has been used. In a similar manner we can show that

$$[\psi(\mathbf{x}, t), \psi(\mathbf{x}', t)]_\pm = [\psi^+(\mathbf{x}, t), \psi^+(\mathbf{x}', t)]_\pm = 0 \qquad (4\text{-}27)$$

We may use these relations to show that the Heisenberg equations of motion

$$-\frac{\hbar}{i}\frac{\partial}{\partial t}\psi(\mathbf{x}', t) = [\psi(\mathbf{x}', t), H]_- \qquad (4\text{-}28)$$

give the time-dependent Schrödinger equation, Eq. 4-1, when the Hamiltonian operator is taken to be

$$H = \int d^3x\psi^+(\mathbf{x}, t)\left[-\frac{\hbar^2}{2m}\nabla^2 + V\right]\psi(\mathbf{x}, t) \qquad (4\text{-}29)$$

Problem 4-1. Show that Eq. 4-1 follows from Eq. 4-28.

We may interpret

$$n(\mathbf{x}, t) = \psi^+(\mathbf{x}, t)\psi(\mathbf{x}, t) \qquad (4\text{-}30)$$

as the number density operator and

$$N(t) = \int d^3x\psi^+(\mathbf{x}, t)\psi(\mathbf{x}, t) \qquad (4\text{-}31)$$

as the total number operator.

Problem 4-2. Show that

$$\frac{dN}{dt} = -\frac{i}{\hbar}[N, H]_- = 0 \qquad (4\text{-}32)$$

so the theory conserves particles.

We shall let $|0\rangle$ denote the vacuum state, that is, the state with no particles present. (It should not be confused with the null vector for which we have previously used the same symbol $|0\rangle$.) Using Eqs. 4-3 and 4-23a or 4-24a we see that

$$\psi(\mathbf{x}, t)\,|0\rangle = 0 \tag{4-33}$$

Now, if our interpretation is correct $\psi^+(\mathbf{x}, t)\,|0\rangle$ should be a state in which there is one particle at $\mathbf{x}$. If we operate on this state with $n(\mathbf{x}', t)$ we find

$$\begin{aligned} n(\mathbf{x}', t)\psi^+(\mathbf{x}, t)\,|0\rangle &= \psi^+(\mathbf{x}', t)\psi(\mathbf{x}', t)\psi^+(\mathbf{x}, t)\,|0\rangle \\ &= \psi^+(\mathbf{x}', t)[\delta(\mathbf{x} - \mathbf{x}') \pm \psi^+(\mathbf{x}, t)\psi(\mathbf{x}', t)]\,|0\rangle \\ &= \delta(\mathbf{x} - \mathbf{x}')\psi^+(\mathbf{x}, t)\,|0\rangle \end{aligned} \tag{4-34}$$

We see that this state is an eigenstate of the number density operator $n(\mathbf{x}', t)$ with an eigenvalue which is zero except at $\mathbf{x}' = \mathbf{x}$ where it is infinite. We also find

$$N\psi^+(\mathbf{x}, t)\,|0\rangle = \psi^+(\mathbf{x}, t)\,|0\rangle \tag{4-35}$$

so the state is an eigenstate of N with eigenvalue unity. This confirms our interpretation of the state $\psi^+(\mathbf{x}, t)\,|0\rangle$.

In a similar way $\psi^+(\mathbf{x}_1, t)\psi^+(\mathbf{x}_2, t)\,|0\rangle$ will be a two-particle state with one particle at $\mathbf{x}_1$ and one at $\mathbf{x}_2$. We can continue and construct states with any number of particles.

Now consider the one-particle state

$$|C_1, t\rangle = \int d^3x\, C_1(\mathbf{x})\psi^+(\mathbf{x}, t)\,|0\rangle \tag{4-36}$$

where $C_1(\mathbf{x})$ is an ordinary function of $\mathbf{x}$ (not an operator). By the usual rules of quantum mechanics the coefficient of $\psi^+(\mathbf{x}, t)\,|0\rangle$ is the probability amplitude for finding the system in the state $\psi^+(\mathbf{x}, t)\,|0\rangle$. Therefore, we interpret

$$|C_1(\mathbf{x})|^2\, d^3x$$

as the probability of finding a particle in d^3x; $|C_1(\mathbf{x})|^2$ plays the same role here that $|\psi(\mathbf{x}, t)|^2$ plays in elementary quantum mechanics. Let us try to choose $C_1(\mathbf{x})$ so that $|C_1, t\rangle$ is an eigenvector of H with the eigenvalue E. That is

$$H\,|C_1, t\rangle = E\,|C_1, t\rangle \tag{4-37}$$

or

$$\begin{aligned} \int d^3x\,\psi^+(\mathbf{x}, t)\left[-\frac{\hbar^2}{2m}\nabla^2 + V\right]\psi(\mathbf{x}, t)\int d^3x'\, C_1(\mathbf{x}')\psi^+(\mathbf{x}', t)\,|0\rangle \\ = E\int d^3x'\, C_1(\mathbf{x}')\psi^+(\mathbf{x}', t)\,|0\rangle \end{aligned} \tag{4-38}$$

We take $\psi(\mathbf{x}, t)$ inside the x'-integral on the left-hand side and use

$$\psi(\mathbf{x}, t)\psi^+(\mathbf{x}', t)\,|0\rangle = [\delta(\mathbf{x} - \mathbf{x}') \pm \psi^+(\mathbf{x}', t)\psi(\mathbf{x}, t)]\,|0\rangle = \delta(\mathbf{x} - \mathbf{x}')\,|0\rangle \tag{4-39}$$

The left-hand side becomes

$$\int d^3x \int d^3x' \psi^+(\mathbf{x}, t)\left[-\frac{\hbar^2}{2m}\nabla^2 + V\right]\delta(\mathbf{x} - \mathbf{x}')C_1(\mathbf{x}')\,|0\rangle$$

$$= \int d^3x \psi^+(\mathbf{x}, t)\left[-\frac{\hbar^2}{2m}\nabla_x{}^2 + V(\mathbf{x})\right]C_1(\mathbf{x})\,|0\rangle \tag{4-40}$$

Combining this with the right-hand side of Eq. 4-38 gives

$$\int d^3x \psi^+(\mathbf{x}, t)\left\{\left[-\frac{\hbar^2}{2m}\nabla_x{}^2 + V(\mathbf{x})\right]C_1(\mathbf{x}) - EC_1(\mathbf{x})\right\}|0\rangle \tag{4-41}$$

It follows that $C_1(\mathbf{x})$ satisfies

$$\left[-\frac{\hbar^2}{2m}\nabla^2 + V\right]C_1(\mathbf{x}) = EC_1(\mathbf{x}) \tag{4-42}$$

This, together with the interpretation of $|C_1|^2$ as the probability density leads us to identify $C_1(\mathbf{x})$ with the single particle wave function of elementary quantum mechanics.

We can construct an n-particle state

$$|C_n, t\rangle = \int d^3x_1 \cdots \int d^3x_n C_n(\mathbf{x}_1 \cdots \mathbf{x}_n)\psi^+(\mathbf{x}, t) \cdots \psi^+(\mathbf{x}_n, t)\,|0\rangle \tag{4-43}$$

We interpret

$$|C_n(\mathbf{x}_1 \cdots \mathbf{x}_n)|^2\, d^3x_1 \cdots d^3x_n$$

as the probability of finding particle 1 in d^3x_1, particle 2 in d^3x_2, and so on. The requirement that $C_n(\mathbf{x}_1 \cdots \mathbf{x}_n)$ be chosen so that

$$H\,|C_n, t\rangle = E\,|C_n, t\rangle \tag{4-44}$$

leads by a straightforward but slightly tedious calculation to the n-particle Schrödinger equation

$$\sum_{i=1}^{n}\left[-\frac{\hbar^2}{2m}\nabla_i{}^2 + V(\mathbf{x}_i)\right]C_n(\mathbf{x}_1 \cdots \mathbf{x}_n) = EC_n(\mathbf{x}_1 \cdots \mathbf{x}_n) \tag{4-45}$$

We see that there is contained within the second quantization formalism the elementary quantum mechanics of an arbitrary number of noninteracting particles.

5

Interaction of Quantized Fields

One can add together the Hamiltonians for several free particle fields and introduce appropriate interaction terms to get a theory of interacting fields, or, equivalently, interacting systems of particles. The interaction we know most about, of course, is the interaction of photons with charged particles. We consider this first, taking the particle field to be described by the Hamiltonian of Eq. 4-29, and the electromagnetic field to be described by the Hamiltonian of Eq. 2-12. The interaction is obtained by the prescription

$$\frac{\hbar}{i}\nabla \to \frac{\hbar}{i}\nabla - \frac{e}{c}\mathbf{A}(\mathbf{x}) \tag{5-1}$$

Making this replacement in Eq. 4-29 and adding the Hamiltonians give the total Hamiltonian

$$\begin{aligned} H &= \int d^3x\psi^+(\mathbf{x}, t)\left[\frac{1}{2m}\left|\frac{\hbar}{i}\nabla - \frac{e}{c}\mathbf{A}\right|^2 + V\right]\psi(\mathbf{x}, t) + \int d^3x\frac{1}{8\pi}(E^2 + B^2) \\ &= H_p + H_{\text{rad}} + H_I \end{aligned} \tag{5-2}$$

where

$$H_p = \int d^3x\psi^+\left[-\frac{\hbar^2}{2m}\nabla^2 + V\right]\psi = \sum_n E_n b_n^+ b_n \tag{5-3}$$

is the particle Hamiltonian,

$$H_{\text{rad}} = \int d^3x\frac{1}{8\pi}(E^2 + B^2) = \sum_{\mathbf{k},\sigma} \hbar\omega_k a_{\mathbf{k}\sigma}^+ a_{\mathbf{k}\sigma} \tag{5-4}$$

is the Hamiltonian of the radiation field, and

$$H_I = \int d^3x\psi^+\left[-\frac{e\hbar}{imc}\mathbf{A}\cdot\nabla + \frac{e^2}{2mc^2}A^2\right]\psi \tag{5-5}$$

is the interaction Hamiltonian. As before, we can divide H_I into a part H' proportional to A and a part H'' proportional to A^2. Expanding $\mathbf{A}$ and ψ in terms of $a_{\mathbf{k}\sigma}$ and b_n gives

$$H_I = H' + H'' \tag{5-6a}$$

$$H' = \sum_{\mathbf{k},\sigma}\sum_{n}\sum_{n'}\{M(\mathbf{k}, \sigma, n, n')b_n^+ b_{n'} a_{\mathbf{k}\sigma} + M(-\mathbf{k}, \sigma, n, n')b_n^+ b_{n'} a_{\mathbf{k},\sigma}^+\} \tag{5-6b}$$

$$\begin{aligned} H'' = \sum_{\mathbf{k}_1\sigma_1}\sum_{k_2\sigma_2}\sum_{n}\sum_{n'} b_n b_{n'}\{&M(k_1, \sigma_1, \mathbf{k}_2, \sigma_2, n, n')a_{\mathbf{k}_1\sigma_1}a_{\mathbf{k}_2\sigma_2} \\ &+ M(\mathbf{k}_1, \sigma_1, -k_2, \sigma_2, n, n')a_{\mathbf{k}_1\sigma_1}a_{\mathbf{k}_2\sigma_2}^+ \\ &+ M(-\mathbf{k}_1, \sigma_1, \mathbf{k}_2, \sigma_2, n, n')a_{\mathbf{k}_1\sigma_1}^+ a_{\mathbf{k}_2\sigma_2} \\ &+ M(-\mathbf{k}_1, \sigma_1, -\mathbf{k}_2, \sigma_2, n, n')a_{\mathbf{k}_1\sigma_1}^+ a_{\mathbf{k}_2\sigma_2}^+\} \end{aligned} \tag{5-6c}$$

where

$$M(\mathbf{k}, \sigma, n, n') = \left(\frac{2\pi\hbar c^2}{\Omega\omega_k}\right)^{1/2} \int d^3x\psi_n^*\left[-\frac{e\hbar}{imc}\, e^{i\mathbf{k}\cdot\mathbf{x}}\mathbf{u}_{k\sigma}\cdot\nabla\right]\psi_{n'} \tag{5-6d}$$

and

$$\begin{aligned} &M(\mathbf{k}_1, \sigma_1, \mathbf{k}_2, \sigma_2, n, n') \\ &\quad = \left(\frac{2\pi\hbar c^2}{\Omega}\right)\frac{1}{(\omega_{k_1}\omega_{k_2})^{1/2}} \int d^3x\psi_n^*\left[\frac{e^2}{2mc^2}\,\mathbf{u}_{k_1\sigma_1}\cdot\mathbf{u}_{k_2\sigma_2}e^{i(\mathbf{k}_1+\mathbf{k}_2)\cdot\mathbf{x}}\right]\psi_n \end{aligned} \tag{5-6e}$$

The part of the Hamiltonian $H_p + H_{\text{rad}}$ may be considered the unperturbed part with eigenvectors

$$|\cdots N_n\cdots\rangle_p\,|\cdots n_{\mathbf{k}\sigma}\cdots\rangle_{\text{rad}} \tag{5-7a}$$

and eigenvalues

$$\sum_n E_n N_n + \sum_{k\sigma}\hbar\omega_k n_{\mathbf{k}\sigma} \tag{5-7b}$$

The interaction Hamiltonian induces transitions between these states. For instance the term $b_n^+ b_{n'} a_{\mathbf{k}\sigma}$ in H' destroys a photon of momentum $\hbar\mathbf{k}$ and polarization σ, destroys a particle in state $|n'\rangle$, and creates a particle in state $|n\rangle$. Such a process can be represented by a diagram like that in Fig. 5-1. In this figure we also draw the diagram that represents the term $b_n^+ b_{n'} a_{\mathbf{k}\sigma}^+$.

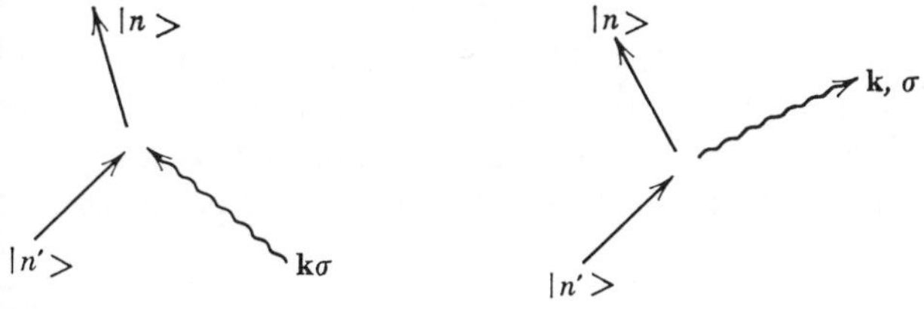

Figure 5-1

The terms in H'' induce transitions in which a particle in state $|n'\rangle$ is destroyed, a particle in state $|n\rangle$ is created, and two photons are destroyed, or one is destroyed and another created, or two are created.

Problem 5-1. Draw diagrams corresponding to the terms in H''.

Problem 5-2. By relabeling indices show that H' and H'' can be put into the manifestly Hermitian forms

$$H' = \sum_{\mathbf{k},\sigma}\sum_{n}\sum_{n'}\{M(\mathbf{k}, \sigma, n, n')b_n^+ b_{n'} a_{\mathbf{k}\sigma} + \text{HC}\} \tag{5-8a}$$

$$H'' = \sum_{\mathbf{k}_1\sigma_1}\sum_{\mathbf{k}_2\sigma_2}\sum_{n}\sum_{n'}\{M(\mathbf{k}_1, \sigma_1, \mathbf{k}_2, \sigma_2, n, n')b_n^+ b_{n'} a_{\mathbf{k}_1\sigma_1} a_{\mathbf{k}_2\sigma_2}$$
$$+ M(\mathbf{k}_1, \sigma_1, -\mathbf{k}_2, \sigma_2, n, n')b_n^+ b_{n'} a_{\mathbf{k}_1\sigma_1} a^+_{\mathbf{k}_2\sigma_2} + \text{HC}\} \tag{5-8b}$$

where HC denotes the Hermitian conjugate of the terms which precede it.

This new formalism has not added any new physics to that which was covered in Chapter 3. Only the way of looking at things is new. For instance, compare the first term in Eq. 5-6b with the first term of Eq. 3-5a. Both terms destroy a photon of momentum $\hbar\mathbf{k}$ and polarization σ. In calculating a transition probability using Eq. 3-5a the operator $\mathbf{p}\cdot\mathbf{u}_{\mathbf{k}\sigma}$ will appear in a scalar product between two atomic states. The same operator occurs between ψ_n^* and $\psi_{n'}$ in Eq. 5-6d. One readily checks that the transition probabilities calculated in either formalism are the same.

Problem 5-3. Repeat enough of the calculations of Chapter 3 using the formalism of the present chapter to convince yourself that the results are the same.

NONRELATIVISTIC BREMSSTRAHLUNG

We have reserved one problem, which could have been treated by the methods of Chapter 3, in order to illustrate the application of the second quantization formalism—the problem of bremsstrahlung.

Classically, an accelerated charged particle radiates. In a collision between two charged particles the particles are accelerated and hence radiate. It is this process of radiation during a collision that we now wish to discuss quantum mechanically. We discuss the collision between an electron and a nucleus, which because of its large mass may be considered fixed. For the purpose of this calculation it is convenient to take the potential V in Eq. 5-2 to be the potential of this scattering nucleus and to treat it as a perturbation. We let

$$H''' = \int d^3x\,\psi^+ V \psi \tag{5-9}$$

The states $|n\rangle$ are now the free particle states which we now denote by $|\mathbf{q}\rangle$, where

$$\psi_{\mathbf{q}}(\mathbf{x}) = \langle \mathbf{x} \mid \mathbf{q} \rangle = \frac{1}{\sqrt{\Omega}} e^{i\mathbf{q}\cdot\mathbf{x}} \tag{5-10}$$

and $\hbar\mathbf{q}$ is the momentum of a particle. Writing

$$\psi(\mathbf{x}, t) = \sum_{\mathbf{q}} b_{\mathbf{q}} \psi_{\mathbf{q}}(\mathbf{x}) \tag{5-11}$$

we find that Eq. 5-3 takes the form

$$H_p = \sum_{\mathbf{q}} \frac{\hbar^2 q^2}{2m} b_{\mathbf{q}}^+ b_{\mathbf{q}} \tag{5-12}$$

The H''' becomes

$$H''' = \sum_{\mathbf{q}} \sum_{\mathbf{q}} b_{\mathbf{q}}^+ b_{\mathbf{q}'} \bar{V}(\mathbf{q} - \mathbf{q}') \tag{5-13a}$$

where

$$\bar{V}(\mathbf{q} - \mathbf{q}') = \int \frac{d^3x}{\Omega} e^{i(\mathbf{q}-\mathbf{q}')\cdot\mathbf{x}} V(\mathbf{x}) \tag{5-13b}$$

is the Fourier transform of $V(\mathbf{x})$.

The integrals in Eqs. 5-6d and 5-6e are easily evaluated. For instance, Eq. 5-6d becomes

$$\begin{aligned} M(k, \sigma, \mathbf{q}, \mathbf{q}') &= \left(\frac{2\pi\hbar c^2}{\Omega\omega_k}\right)^{1/2} \left(-\frac{e\hbar}{mc}\right) (\mathbf{q}' \cdot \mathbf{u}_{\mathbf{k}\sigma}) \int \frac{d^3x}{\Omega} e^{i(\mathbf{q}'+\mathbf{k}-\mathbf{q})\cdot\mathbf{x}} \\ &= \left(\frac{2\pi\hbar c^2}{\Omega\omega_k}\right)^{1/2} \left(-\frac{e\hbar}{mc}\right) (\mathbf{q} \cdot \mathbf{u}_{\mathbf{k}\sigma}) \delta_{\mathbf{q},\mathbf{q}'+\mathbf{k}} \end{aligned} \tag{5-14}$$

The H' becomes

$$H' = -\frac{e\hbar}{mc} \sum_{\mathbf{k},\sigma} \sum_{\mathbf{q}} \left(\frac{2\pi\hbar c^2}{\Omega\omega_k}\right)^{1/2} \mathbf{q} \cdot \mathbf{u}_{\mathbf{k}\sigma} \{ b_{\mathbf{q}+\mathbf{k}}^+ b_{\mathbf{q}} a_{\mathbf{k}\sigma} + b_{\mathbf{q}}^+ b_{\mathbf{q}+\mathbf{k}} a_{\mathbf{k}\sigma}^+ \} \tag{5-15}$$

We can consider Bremsstrahlung as a second-order process in which both H' and H''' are treated as perturbations. The second-order term in Eq. 1-166 must be used in this calculation. In our present notation this is

$$M_{fi} = \sum_I \frac{\langle f | \, H' + H''' \, | I \rangle \langle I | \, H' + H''' \, | i \rangle}{E_i - E_I + i\eta} \tag{5-16}$$

We can describe the process by the diagrams of Fig. 5-2. We have indicated

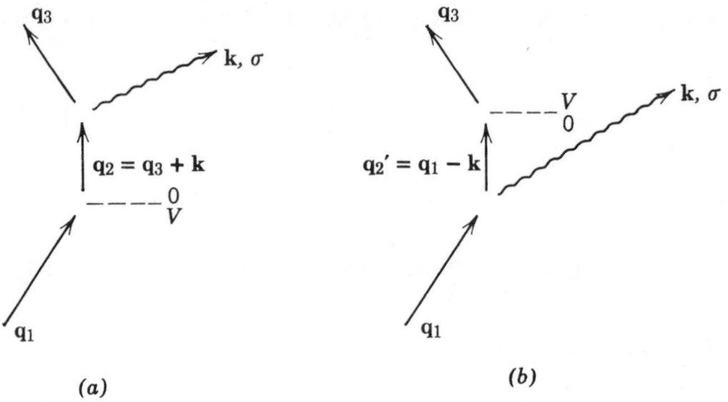

Figure 5-2

the action of H''' by the dotted line and V. The initial and final states are

$$|i\rangle = |\text{one electron with } \mathbf{q}_1\rangle_p \, |\text{no photons}\rangle_{\text{rad}} \tag{5-17a}$$

$$|f\rangle = |\text{one electron with } q_3\rangle_p \, |\text{one photon with } \mathbf{k}, \sigma\rangle_{\text{rad}} \tag{5-17b}$$

The transition from $|i\rangle$ to $|f\rangle$ can take place through either of the intermediate states

$$|I_1\rangle = |\text{one electron with } \mathbf{q}_2\rangle_p \, |\text{no photons}\rangle_{\text{rad}} \tag{5-18a}$$

$$|I_2\rangle = |\text{one electron with } \mathbf{q}_2'\rangle_p \, |\text{one photon with } \mathbf{k}, \boldsymbol{\sigma}\rangle_{\text{rad}} \tag{5-18b}$$

These two ways of reaching $|f\rangle$ from $|i\rangle$ are shown in Figs. 5-2*a* and 5-2*b*. Conservation of momentum at the vertex where the photon is emitted gives

$$\mathbf{q}_2 = \mathbf{q}_3 + \mathbf{k} \tag{5-19a}$$

and

$$\mathbf{q}_2' = \mathbf{q}_1 - \mathbf{k} \tag{5-19b}$$

Equation 5-16 becomes

$$M_{fi} = \frac{\langle f|\, H' \,|I_1\rangle\langle I_1|\, H''' \,|i\rangle}{E_i - E_{I_1}} + \frac{\langle f|\, H''' \,|I_2\rangle\langle I_2|\, H' \,|i\rangle}{E_i - E_{I_2}} \tag{5-20}$$

(The infinitesimal η is not needed here.)

For the Coulomb potential

$$V(x) = \frac{-Ze^2}{r} \tag{5-21a}$$

Equation 5-13b gives

$$\bar{V}(\mathbf{q}) = -\frac{4\pi Ze^2}{\Omega q^2} \tag{5-21b}$$

The matrix elements and energies which enter Eq. 5-20 are

$$\langle I_1|\, H''' \,|i\rangle = \langle f|\, H''' \,|I_2\rangle = -\frac{4\pi Ze^2}{\Omega\, |\mathbf{q}_1 - \mathbf{q}_3 - \mathbf{k}|^2} \tag{5-22a}$$

$$\langle f\,|\, H' \,|I_1\rangle = -\frac{e\hbar}{mc}\left(\frac{2\pi\hbar c^2}{\Omega\omega_k}\right)^{1/2} \mathbf{q}_3 \cdot \mathbf{u}_{\mathbf{k}\sigma} \tag{5-22b}$$

$$\langle I_2|\, H' \,|i\rangle = -\frac{e\hbar}{mc}\left(\frac{2\pi\hbar c^2}{\Omega\omega_k}\right)^{1/2} \mathbf{q}_1 \cdot \mathbf{u}_{\mathbf{k}\sigma} \tag{5-22c}$$

$$E_i = \frac{\hbar^2 q_1^{\,2}}{2m} \tag{5-22d}$$

$$E_{I_1} = \frac{\hbar^2}{2m}\, |\mathbf{q}_3 + \mathbf{k}|^2 \tag{5-22e}$$

$$E_{I_2} = \frac{\hbar^2}{2m}\, |\mathbf{q}_1 - \mathbf{k}|^2 + \hbar\omega_k \tag{5-22f}$$

The final energy is

$$E_f = \frac{\hbar^2}{2m}\, q_3^{\,2} + \hbar\omega_k \tag{5-23}$$

By conservation of energy this must equal E_i. This may be used to simplify the energy denominators in Eq. 5-20:

$$(E_i - E_{I_1}) = \hbar\omega_k\left(1 - \frac{\mathbf{k}\cdot\mathbf{v}_3}{kc} - \frac{\hbar k}{2mc}\right) \tag{5-24a}$$

$$(E_i - E_{I_2}) = -\hbar\omega_k\left(1 - \frac{\mathbf{k}\cdot\mathbf{v}_1}{kc} + \frac{\hbar k}{2mc}\right) \tag{5-24b}$$

where we have introduced the velocities $\mathbf{v}_1 = \hbar q_1/m$ and $\mathbf{v}_3 = \hbar\mathbf{q}_3/m$.

Equation 5-20 becomes

$$M_{fi} = \left(\frac{4\pi Ze^2}{\Omega\, |\mathbf{q}_1 - \mathbf{q}_3 - \mathbf{k}|^2}\right)\left(\frac{2\pi e^2}{\hbar\Omega\omega_k^{\,3}}\right)^{1/2}$$
$$\times\, \mathbf{u}_{\mathbf{k}} \cdot \left(\frac{\mathbf{v}_3}{1 - \mathbf{k}\cdot(\mathbf{v}_3/kc) - (\hbar k/2mc)} - \frac{\mathbf{v}_1}{1 - \mathbf{k}\cdot(\mathbf{v}_1/kc) + (\hbar k/2mc)}\right) \tag{5-25}$$

If we make the approximation that the electrons are nonrelativistic so that $v/c \ll 1$ and assume that the momentum of the photon $\hbar k$ is much less than the particles momentum, then Eq. 5-25 simplifies to

$$M_{fi} = \left(\frac{4\pi Ze^2\hbar^2}{\Omega m^2\, |\Delta\mathbf{v}|^2}\right)\left(\frac{2\pi e^2}{\hbar\Omega\omega^3}\right)^{1/2} \mathbf{u}\cdot\Delta\mathbf{v} \tag{5-26}$$

where $\Delta\mathbf{v} = \mathbf{v}_3 - \mathbf{v}_1$, and we have dropped the subscripts on ω_k and $\mathbf{u}_{k\sigma}$.

To calculate the total cross section for the process we must sum the transition probability over all final states and equate the result to the product of the cross section and the flux of electrons which is v/Ω.

Thus

$$\left(\frac{v}{\Omega}\right)\sigma = \frac{\Omega^2}{(2\pi)^6}\int q_3{}^2\,dq_3\,d\Omega_e \int k^2\,dk\,d\Omega_k \frac{2\pi}{\hbar}|M_{fi}|^2\,\delta\left[\frac{\hbar^2 q_3{}^2}{2m} - \frac{\hbar^2 q_1{}^2}{2m}\right] \tag{5-27}$$

where we have used Eq. 3-12 once for the final state of the electron and again for the final state of the photon. We have let $d\Omega_e$ be the element of solid angle into which the electron is scattered and $d\Omega_k$ be the element of solid angle into which the photon is scattered. Using Eq. 5-26 for M_{fi} and $|\Delta\mathbf{v}| = 2v\sin\theta/2$ we find

$$\sigma = \left(\frac{\pi Z^2 e^4}{m^2\,|\Delta\mathbf{v}|^4}\right)\left(\frac{e^2}{4\pi^2 c^3\hbar}\right)\int d\Omega_e \int d\Omega_k \int \frac{d\omega}{\omega}\frac{|\mathbf{u}\cdot\Delta\mathbf{v}|^2}{\sin^4\theta/2} \tag{5-28}$$

We may interpret

$$\frac{d^3\sigma}{d\Omega_e\,d\Omega_k\,d\omega} = \left(\frac{\pi Z^2 e^4}{m^2 v^4 \sin^4\theta/2}\right)\left(\frac{e^2\,|\mathbf{u}\cdot\Delta\mathbf{v}|^2}{4\pi^2 c^3\hbar\omega}\right) \tag{5-29}$$

as the cross section for scattering an electron through the angle θ into the element of solid angle $d\Omega_e$ with the emission of a photon with frequency ω in the range $d\omega$ and polarization $\mathbf{u}$ into the solid angle $d\Omega_k$.

The first factor in Eq. 5-29 will be recognized as the Rutherford cross section for electron scattering. The second factor gives the probability that a photon of frequency ω and polarization $\mathbf{u}$ is emitted into $d\omega\,d\Omega_k$.

6

Quantum Electrodynamics

To arrive at a more satisfactory theory of charged particles, photons, and their interaction, two modifications must be made in the formalism developed in the preceding chapters. First, the electrons should be treated by the Dirac equation rather than the nonrelativistic Schrödinger equation. Second, the entire electromagnetic field, rather than just the transverse part, should be treated quantum mechanically.

The first of these objectives is accomplished by replacing the particle Hamiltonian of Eq. 5-3 by

$$H_p = \int d^3x \psi^+ \left(\frac{\hbar c}{i} \boldsymbol{\alpha} \cdot \nabla + \beta mc^2 \right) \psi \tag{6-1}$$

where

$$\psi = \begin{bmatrix} \psi_1 \\ \psi_2 \\ \psi_3 \\ \psi_4 \end{bmatrix} \quad \text{and} \quad \psi^+ = [\psi_1^+, \psi_2^+, \psi_3^+, \psi_4^+] \tag{6-2}$$

and $\boldsymbol{\alpha}$ and β are the Dirac matrices given in Eq. A-25 of the Appendix. The components of ψ and ψ^+ are now considered to be operators. We assume that

$$\begin{aligned} [\psi_j(\mathbf{x}, t), \psi_k(\mathbf{x}', t)]_+ &= [\psi_j^+(\mathbf{x}, t), \psi_k^+(\mathbf{x}', t)]_+ = 0 \\ [\psi_j(\mathbf{x}, t), \psi_k^+(\mathbf{x}', t)]_+ &= \delta_{jk}\, \delta(\mathbf{x} - \mathbf{x}') \end{aligned} \tag{6-3}$$

in analogy with Eqs. 4-26 and 4-27. We have chosen the anticommutation relations, since we are developing a theory of particles that obey Fermi-Dirac statistics. It is straightforward to show that the Heisenberg equations of motion

$$-\frac{\hbar}{i}\frac{\partial\psi}{\partial t} = [\psi, H_p]_- \tag{6-4}$$

lead to the Dirac equation

$$-\frac{\hbar}{i}\frac{\partial\psi}{\partial t} = \left(\frac{\hbar c}{i}\boldsymbol{\alpha}\cdot\boldsymbol{\nabla} + \beta mc^2\right)\psi \tag{6-5}$$

As shown in the Appendix, the Dirac Hamiltonian has the eigenfunctions

$$\psi_{\mathbf{p},\lambda} = u_{\mathbf{p},\lambda}\frac{e^{i\mathbf{p}\cdot\mathbf{x}}}{\sqrt{\Omega}} \tag{6-6}$$

where $\lambda = 1, 2, 3, 4$. The energy eigenvalues are

$$E_{\mathbf{p},\lambda} = \pm\sqrt{\hbar^2c^2p^2 + m^2c^4} \tag{6-7}$$

where the plus sign is to be taken for $\lambda = 1, 2$ and the minus sign is to be taken for $\lambda = 3, 4$. One choice of the four component Dirac spinors $u_{\mathbf{p},\lambda}$ is given in Eq. A-55.

In analogy with our procedure in Chapter 4, we expand ψ and ψ^+ as

$$\psi(\mathbf{x}, t) = \sum_{\mathbf{p},\lambda} b_{\mathbf{p},\lambda}(t)\psi_{\mathbf{p},\lambda}(\mathbf{x}) \tag{6-8}$$

and

$$\psi^+(\mathbf{x}, t) = \sum_{\mathbf{p},\lambda} b^+_{\mathbf{p}\lambda}(t)\psi^+_{\mathbf{p},\lambda}(\mathbf{x}) \tag{6-9}$$

Substitution into Eq. 6-5 gives

$$\frac{d}{dt}b_{\mathbf{p},\lambda} = -\frac{i}{\hbar}E_{\mathbf{p}\lambda}b_{\mathbf{p}\lambda} \tag{6-10}$$

Equation 6-1 becomes

$$H_p = \sum_{\mathbf{p}\lambda} E_{\mathbf{p}\lambda}b^+_{\mathbf{p}\lambda}b_{\mathbf{p}\lambda} \tag{6-11}$$

It is easily shown that the Heisenberg equations of motion

$$-\frac{\hbar}{i}\frac{d}{dt}b_{\mathbf{p}\lambda} = [b_{\mathbf{p}\lambda}, H_p]_- \tag{6-12}$$

yield Eq. 6-10 when H_p is given by Eq. 6-11.

The interaction of the particles and the electromagnetic field is obtained by the usual prescription of replacing $\mathbf{p}$ by $\mathbf{p} - e\mathbf{A}/c$. If we do this in Eq. 6-1 we find the interaction Hamiltonian

$$H_I = -e\int d^3x\psi^+\boldsymbol{\alpha}\cdot\mathbf{A}\psi \tag{6-13}$$

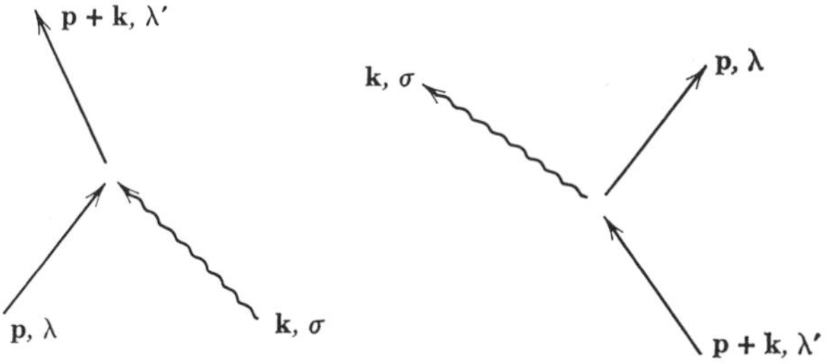

(a) (b)

Figure 6-1

Expanding ψ, ψ^+, and $\mathbf{A}$ gives

$$H_I = -e \sum_{\mathbf{k}\sigma} \sum_{\mathbf{p}\lambda} \sum_{\lambda'} \left\{ \left(\frac{2\pi\hbar c^2}{\Omega\omega_k} \right)^{1/2} \{ (u^+_{\mathbf{p}+\mathbf{k},\lambda'} \boldsymbol{\alpha} \cdot \mathbf{u}_{\mathbf{k}\sigma} u_{\mathbf{p}\lambda}) b^+_{\mathbf{p}+\mathbf{k},\lambda'} b_{\mathbf{p},\lambda} a_{\mathbf{k}\sigma} + \mathrm{HC} \} \right\} \qquad (6\text{-}14)$$

where HC stands for the Hermitian conjugate of the preceding term. The terms in H_I may be represented by the Feynman diagrams of Fig. 6-1.

In Fig. 6-1*a* the operator $b_{\mathbf{p}\lambda}$ destroys an electron of momentum $\hbar\mathbf{p}$, the operator $a_{\mathbf{k}\sigma}$ destroys a photon of momentum $\hbar\mathbf{k}$, and the operator $b^+_{\mathbf{p}+\mathbf{k},\lambda'}$ creates an electron of momentum $\hbar(\mathbf{p} + \mathbf{k})$. (Momentum conservation came from the spatial integration just as it did in the derivation of Eq. 5-15.) The Hermitian conjugate of $b^+_{\mathbf{p}+\mathbf{k},\lambda'}\, b_{\mathbf{p}\lambda} a_{\mathbf{k}\sigma}$ is the operator $a^+_{\mathbf{k}\sigma} b^+_{\mathbf{p}\lambda} b_{\mathbf{p}+\mathbf{k},\lambda'}$ which is represented by the diagram of Fig. 6-1*b*. It should be noted that the quantum number λ can change when a photon is emitted or absorbed. This indicates that the spin and the sign of the energy can be changed in the process.

Next we consider the second modification that must be made in the theory: the inclusion of the entire electromagnetic field in the formalism rather than just its transverse part. In working in the Coulomb gauge, the part of the electromagnetic field that gives the Coulomb interaction between particles (i.e., such terms as $e_i e_j / r_{ij}$ in the Hamiltonian) was not derivable from the vector potential $\mathbf{A}$. We can remedy this defect and at the same time make the theory manifestly covariant by replacing the three-vector $\mathbf{A}$ by the four-vector potential which we denote by A_μ. The fourth component of A_μ is the scalar potential ϕ, and the electric and magnetic fields are given by Eqs. 2-1. Equation 2-11a must be changed to

$$A_\mu = \sum_{\mathbf{k}} \sum_{\sigma=1}^{4} \left(\frac{2\pi\hbar c^2}{\Omega\omega_k} \right)^{1/2} u_{\mathbf{k}\sigma\mu} \{ a_{\mathbf{k}\sigma} e^{i\mathbf{k}\cdot\mathbf{x}} + a^+_{\mathbf{k}\sigma} e^{-i\mathbf{k}\cdot\mathbf{x}} \} \qquad (6\text{-}15)$$

The polarization vectors $\mathbf{u}_{\mathbf{k}\sigma}$ have been replaced by four-vectors $u_{\mathbf{k}\sigma\mu}$. Also, the index σ now takes on the values $\sigma = 1, 2, 3, 4$ instead of only $\sigma = 1, 2$.

This allows for the possibility of "longitudinal" photons (polarized along the direction of **k**) and "time-like" photons (with their polarization vector along the time axis).

These exotic photons are needed to describe the part of the electromagnetic field previously omitted from our treatment. In this formalism the Coulomb interaction between two charged particles can be described as the emission of a photon by one of the particles and its absorption by the other. The longitudinal and time-line photons never exist as free particles. (However, in a plasma it is possible to have free longitudinal photons. These are called plasmons and are the quanta of the electrostatic oscillations of the plasma.)

Equation 6-13 for the interaction Hamiltonian must be modified to take into account the change of **A** to a 4-vector. This is accomplished if it is recalled that $\psi^+\boldsymbol{\alpha}\psi$ is the three-vector part of a 4-vector whose fourth component is $\psi^+\psi$. If we define

$$\gamma_i = -i\beta\alpha_i \qquad \text{for } i = 1, 2, 3 \tag{6-16a}$$

and

$$\gamma_4 = \beta \tag{6-16b}$$

and let

$$\bar{\psi} = i\psi^+\beta \tag{6-17}$$

then $\bar{\psi}\gamma_\mu\psi$ is a 4-vector. The interaction Hamiltonian can be written as

$$H_I = -e\int d^3x\bar{\psi}\gamma_\mu\psi A_\mu \tag{6-18}$$

The integrand is the scalar product of two 4-vectors, hence is relativistically invariant. Equation 6-18 reduces to Eq. 6-13 in the Coulomb gauge. Equation 6-14 becomes

$$H_I = -e\sum_{\mathbf{k},\sigma}\sum_{\mathbf{p}\lambda}\sum_{\lambda'}\left(\frac{2\pi\hbar c^2}{\Omega\omega_k}\right)^{1/2}\{(\bar{u}_{\mathbf{p}+\mathbf{k},\lambda'}\gamma_\mu u_{\mathbf{k}\sigma\mu}u_{\mathbf{p},\lambda})b^+_{\mathbf{p}+\mathbf{k},\lambda'}b_{\mathbf{p}\lambda}a_{\mathbf{k}\sigma} + \text{HC}\} \tag{6-19}$$

In the applications that we discuss we do not need this more general formalism which includes longitudinal and time-line photons. It is always possible to choose the Coulomb gauge. For our purposes this is the most convenient choice.

DIRAC'S HOLE THEORY

The relativistic theory encounters a difficulty that was not present in the nonrelativistic theory. This is the problem of the negative energy states. One cannot just exclude such states as could be done in a classical theory. Without the negative energy states the eigenfunctions of the Dirac Hamiltonian would not form a complete set. Furthermore, as will be seen when Compton scattering is discussed, the negative energy states are essential in getting the right answer for this well known effect. However, it would appear that electrons in positive energy states would all make radiative transitions into negative energy states and in reality this certainly does not occur.

Dirac overcame this difficulty by making the following two assumptions:

1. In the normal state of the vacuum all negative energy states are occupied and all positive energy states are empty. That is,

$$N_{\mathbf{p},1} = N_{\mathbf{p},2} = 0 \tag{6-20a}$$

$$N_{\mathbf{p},3} = N_{\mathbf{p},4} = 1 \tag{6-20b}$$

where $N_{\mathbf{p},\lambda}$ is the eigenvalue of $b^+_{\mathbf{p}\lambda}b_{\mathbf{p}\lambda}$. Since the negative energy states are full, the exclusion principle forbids transitions into these states. This gets rid of the difficulty just mentioned.

2. This infinite sea of negative energy electrons produces no observable effect.

Now, the eigenvalues of H_p are

$$E = \sum_{\mathbf{p}\lambda} E_{\mathbf{p}\lambda} N_{\mathbf{p}\lambda} \tag{6-21}$$

so that the energy of the vacuum is negatively infinite. This is an additive constant to the energy and will cancel out when energy differences are taken. Therefore it may be disregarded just as the zero-point energy of the radiation field was in Chapter 2.

Now, consider the term $b_{\mathbf{p}+\mathbf{k},\lambda'}b_{\mathbf{p}\lambda}a_{\mathbf{k}\sigma}$ in Eq. 6-14 and suppose that $\lambda = 3$ or 4 and $\lambda' = 1$ or 2. The operator $a_{\mathbf{k}\sigma}$ destroys a photon, the operator $b_{\mathbf{p}\lambda}$ destroys a particle of momentum $\hbar\mathbf{p}$ in the sea of negative energy electrons, and the operator $b^+_{\mathbf{p}+\mathbf{k},\lambda'}$ creates a positive energy electron of momentum $\hbar(\mathbf{p} + \mathbf{k})$. This positive energy electron is observable. Also, the destruction of the negative energy electron has decreased the charge of the universe by $-e$, has decreased the momentum by $\hbar\mathbf{p}$, and has decreased the angular momentum by the spin of the particle that was destroyed. This change in the universe should be observable. It may be thought of as a "hole" in the infinite sea of negative energy electrons. It is equivalent to the creation of a particle of charge $+e$, momentum $-\hbar\mathbf{p}$, and angular momentum opposite to that of the particle destroyed. The process just described should appear as the absorption of a photon together with the production of an electron-position pair. An operator like $b_{\mathbf{p},3}$ which destroys a negative energy electron of momentum $\hbar\mathbf{p}$ and spin of $+\hbar/2$ along the z-axis may also be considered to be the creation operator of a positron of momentum $-\hbar\mathbf{p}$ and spin $-\hbar/2$.

It is amusing to read papers on the quantum theory of radiation (Fermi's paper in *Reviews of Modern Physics*[10] is recommended) written between the discovery of the Dirac equation of 1928 and Anderson's experimental discovery of the positron in 1933. Attempts were made to identify the "hole" with the proton. Reasons were found for believing that the mass of the hole should be greater than the electron mass predicted by the theory. However, it should be possible for an electron and a hole to annihilate with the emission of two photons. When the probability of this process was calculated by Oppenheimer,[31] Dirac,[32] and Tamm[33] it was found that matter would be destroyed in a very short time. When the positron was discovered, what had previously been a major shortcoming of the theory became its greatest triumph. The prediction of the existence of this previously unobserved particle must be regarded as one of the greatest successes of theoretical physics.

ČERENKOV RADIATION BY A DIRAC ELECTRON

Since Čerenkov radiation is a first order process involving free particles it is a particularly simple application of the theory to discuss. Just as in Chapter 3, the interaction Hamiltonian for photons and electrons in a dispersive medium is obtained by replacing ω_k in Eq. 6-14 by

$$\left(\frac{1}{2}\frac{\partial}{\partial\omega}\,\omega^2\varepsilon(\omega)\right)$$

where $\varepsilon(\omega)$ is the dielectric function of the medium.

We consider the process in which an electron of momentum $\hbar(\mathbf{p}+\mathbf{k})$ emits a photon of momentum $\hbar\mathbf{k}$ and polarization σ. Using Eq. 6-14 and first order perturbation theory gives the transition probability per unit time

$$\left(\frac{\text{trans prob}}{\text{time}}\right)_{\mathbf{p}+\mathbf{k},\lambda'\to\mathbf{p},\lambda} = \frac{2\pi}{\hbar}\,e^2\left[\frac{2\pi\hbar c^2}{\Omega\frac{1}{2}(\partial/\partial\omega)\omega^2\varepsilon}\right]|u^+_{\mathbf{p}+\mathbf{k},\lambda'}\boldsymbol{\alpha}\cdot\mathbf{u}_{k\sigma}u_{\mathbf{p}\lambda}|^2 \times\delta[\sqrt{\hbar^2c^2\,|\mathbf{p}+\mathbf{k}|^2+m^2c^4}-\sqrt{\hbar^2c^2p^2+m^2c^4}-\hbar\omega] \quad (6\text{-}22)$$

A little algebra shows that the angle between $\mathbf{p}$ and $\mathbf{k}$ is given by Eq. 3-72.

We may proceed to calculate the energy loss per unit length as we did in Eq. 3-59. There is one modification that we wish to make in this calculation. The sum over final states must include a sum over the final spin states of the electron $\lambda = 1, 2$. Also we shall average over the initial spin states; thus what we want is

$$\frac{dW}{dx} = \frac{1}{v}\frac{1}{2}\sum_{\lambda'=1}^{2}\sum_{\lambda=1}^{2}\sum_{\mathbf{k},\sigma}^{2}\hbar\omega_k\left(\frac{\text{trans prob}}{\text{time}}\right) \tag{6-23}$$

We must evaluate

$$\frac{1}{2}\sum_{\lambda'=1}^{2}\sum_{\lambda=1}^{2}(u^+_{\mathbf{p}+\mathbf{k},\lambda'}\boldsymbol{\alpha}\cdot\mathbf{u}_{\mathbf{k}\sigma}u_{\mathbf{p}\lambda})(u^+_{\mathbf{p}\lambda}\boldsymbol{\alpha}\cdot\mathbf{u}_{\mathbf{k}\sigma}u_{\mathbf{p}+\mathbf{k},\lambda'}) \tag{6-24}$$

If the reader will try to evaluate this straightforwardly he will soon become convinced that there must be some easier way, and indeed there is. The first step is to extend the sums over λ' and λ to include all four values. We can do this by noting that

$$\frac{H_\mathbf{p} + |E_\mathbf{p}|}{2\,|E_\mathbf{p}|}u_{\mathbf{p},\lambda} = \begin{cases} u_{\mathbf{p},\lambda} & \text{for } \lambda = 1, 2 \\ 0 & \text{for } \lambda = 3, 4 \end{cases} \tag{6-25}$$

where

$$H_\mathbf{p} = c\boldsymbol{\alpha}\cdot\mathbf{p} + \beta mc^2 \tag{6-26}$$

We can use this with a similar relation involving $u_{\mathbf{p}+\mathbf{k},\lambda'}$ to write Eq. 6-24 as

$$\frac{1}{2}\sum_{\lambda'=1}^{4}\sum_{\lambda=1}^{4}[u^+_{\mathbf{p}+\mathbf{k},\lambda'}\boldsymbol{\alpha}\cdot\mathbf{u}_{\mathbf{k}\sigma}(H_\mathbf{p} + |E_\mathbf{p}|)u_{\mathbf{p}\lambda}]$$

$$\times\ [u^+_{\mathbf{p}\lambda}\boldsymbol{\alpha}\cdot\mathbf{u}_{\mathbf{k}\sigma}(H_{\mathbf{p}+\mathbf{k}} + |E_{\mathbf{p}+\mathbf{k}}|)u_{\mathbf{p}+\mathbf{k},\lambda'}]\frac{1}{4\,|E_\mathbf{p}|\,|E_{\mathbf{p}+\mathbf{k}}|} \tag{6-27}$$

Now consider

$$\sum_{\lambda=1}^{4}u_{\mathbf{p}\lambda}u^+_{\mathbf{p}\lambda}$$

By the completeness relation this is just the 4×4 unit matrix. Therefore, Eq. 6-27 becomes

$$\frac{1}{2}\sum_{\lambda'=1}^{4}[u^+_{\mathbf{p}+\mathbf{k},\lambda'}\boldsymbol{\alpha}\cdot\mathbf{u}_{k\sigma}(H_\mathbf{p} + |E_\mathbf{p}|)\boldsymbol{\alpha}\cdot\mathbf{u}_{k\sigma}(H_{\mathbf{p}+\mathbf{k}} + |E_{\mathbf{p}+\mathbf{k}}|)u_{\mathbf{p}+\mathbf{k},\lambda'}]\cdot\frac{1}{4\,|E_\mathbf{p}|\,|E_{\mathbf{p}+\mathbf{k}}|}$$

$$= \frac{1}{8\,|E_\mathbf{p}|\,|E_{\mathbf{p}+\mathbf{k}}|}\text{Tr}\,[\boldsymbol{\alpha}\cdot\mathbf{u}_{\mathbf{k}\sigma}(H_\mathbf{p} + |E_\mathbf{p}|)\boldsymbol{\alpha}\cdot\mathbf{u}_{\mathbf{k}\sigma}(H_{\mathbf{p}+\mathbf{k}} + |E_{\mathbf{p}+\mathbf{k}}|)] \tag{6-28}$$

This trace can be evaluated without great difficulty. First we note that

$$\text{Tr}\,\alpha_i = \text{Tr}\,\beta = 0 \tag{6-29}$$

Also, it is easily shown that the trace of a product of any odd number of the matrices α_x, α_y, α_z, and β is zero. We may use the identity

$$(\boldsymbol{\alpha}\cdot\mathbf{a})(\boldsymbol{\alpha}\cdot\mathbf{b}) = 2(\mathbf{a}\cdot\mathbf{b})1 - (\boldsymbol{\alpha}\cdot\mathbf{b})(\boldsymbol{\alpha}\cdot\mathbf{a}) \tag{6-30}$$

where **a** and **b** are arbitrary vectors, and

$$\text{Tr}\,AB = \text{Tr}\,BA \tag{6-31}$$

to show that

$$\mathrm{Tr}\,(\boldsymbol{\alpha}\cdot\mathbf{a})(\boldsymbol{\alpha}\cdot\mathbf{b}) = 4\mathbf{a}\cdot\mathbf{b} \tag{6-32}$$

Also, we can show that

$$\mathrm{Tr}\,(\boldsymbol{\alpha}\cdot\mathbf{a})\beta(\boldsymbol{\alpha}\cdot\mathbf{b})\beta = -4(\mathbf{a}\cdot\mathbf{b}) \tag{6-33}$$

and

$$\mathrm{Tr}\,(\boldsymbol{\alpha}\cdot\mathbf{a})(\boldsymbol{\alpha}\cdot\mathbf{b})(\boldsymbol{\alpha}\cdot\mathbf{c})(\boldsymbol{\alpha}\cdot\mathbf{d}) = 4(\mathbf{a}\cdot\mathbf{b})(\mathbf{c}\cdot\mathbf{d}) - 4(\mathbf{a}\cdot\mathbf{c})(\mathbf{b}\cdot\mathbf{d}) + 4(\mathbf{a}\cdot\mathbf{d})(\mathbf{b}\cdot\mathbf{c}) \tag{6-34}$$

for any four vectors **a**, **b**, **c**, **d**. These relations are very useful in many calculations in quantum electrodynamics.

Problem 6-1. Use Eqs. A-24 to prove Eq. 6-30 and then prove Eqs. 6-32, 6-33, and 6-34.

Using these trace formulas, Eq. 6-28 can be evaluated. We find

$$\frac{1}{2}\left\{1 - \frac{m^2c^4}{|E_{\mathbf{p}}|\,|E_{\mathbf{p}+\mathbf{k}}|} + \frac{2(\mathbf{u}_{k\sigma}\cdot\mathbf{v}_1)^2}{c^2} - \frac{\mathbf{v}_1\cdot\mathbf{v}_2}{c^2}\right\} \tag{6-35}$$

where we have used $\mathbf{v} = c^2\mathbf{p}/E$ and have let $\mathbf{v}_1$ and $\mathbf{v}_2$ be the velocities before and after the emission of the photon. The sum over polarizations can be carried out as was done in Eq. 3-69. The result is that Eq. 6-24 summed over polarizations is

$$\frac{{v_1}^2}{c^2}(1-\cos^2\theta) + \frac{1}{2}\left\{1 - \sqrt{(1-{v_1}^2/c^2)(1-{v_2}^2/c^2)} - \frac{\mathbf{v}_1\cdot\mathbf{v}_2}{c^2}\right\} \tag{6-36}$$

where again θ is the angle between **p** and **k** and is given by Eq. 3-72. We have used

$$E = \frac{mc^2}{\sqrt{1-v^2/c^2}} \tag{6-37}$$

to obtain Eq. 6-36 from Eq. 6-35. The second term in Eq. 6-36 is a small correction to the result we found in Chapter 3. If the momentum of the photon is negligible in comparison with the momentum of the electron then $\mathbf{v}_1 \simeq \mathbf{v}_2$ and the term in braces vanishes. This will be true in both the classical limit ($\hbar \to 0$) and the extreme relativistic limit ($\mathbf{v} \to c$). We neglect this term in the remainder of the calculation. The rest of the calculation parallels that in Chapter 3. The only difference is that Eq. 3-72 must be used for $\cos\theta$ instead of Eq. 3-67. The result is

$$\frac{dW}{dx} = \frac{e^2}{c^2}\int \frac{\varepsilon(\omega)\omega^2\,d\omega}{\frac{1}{2}(\partial/\partial\omega)\omega^2\varepsilon}\left[1 - \frac{c^2}{n^2v^2}\left(1 + \frac{\hbar\omega}{2mc^2}(n^2-1)\sqrt{1-v^2/c^2}\right)^2\right] \tag{6-38}$$

COMPTON SCATTERING

In Chapter 3 we discussed the scattering of a photon by a nonrelativistic electron. We wish to reconsider the problem now using the relativistic theory. In the previous theory the scattering was produced in first order by the A^2 term in the interaction Hamiltonian. There is no A^2 term in Eq. 6-13, so that it is clear that in the relativistic theory scattering must be a second order process. We can picture the scattering process as occurring as shown in the Feynman diagrams of Fig. 6-2. If Fig. 6-2*a* the electron first absorbs a photon of momentum $\hbar\mathbf{k}_i$ and polarization σ_i and then emits a photon of momentum $\hbar\mathbf{k}_f$ and polarization σ_f. If Fig. 6-2*b* the time order of these events is reversed. The initial and final states are

$$|i\rangle = |\mathbf{q}_i, \lambda_i\rangle_e \, |\cdots n_{k_i\sigma_i} \cdots n_{\mathbf{k}_f\sigma_f} \cdots\rangle_{\text{rad}} \tag{6-39a}$$

$$|f\rangle = |\mathbf{q}_f, \lambda_f\rangle_e \, |\cdots n_{k_i\sigma_i} - 1, \cdots n_{k_f\sigma_f} + 1, \cdots\rangle_{\text{rad}} \tag{6-39b}$$

where

$$\langle \mathbf{x} \mid \mathbf{q}, \lambda\rangle = \psi_{\mathbf{q}\lambda}(\mathbf{x}) = u_{\mathbf{q}\lambda} \frac{e^{i\mathbf{q}\cdot\mathbf{x}}}{\sqrt{\Omega}} \tag{6-40}$$

The process occurs through the intermediate states

$$|I_1\rangle = |\mathbf{q}_1, \lambda_1\rangle_e \, |\cdots n_{\mathbf{k}_i\sigma_i} - 1, \cdots n_{k_f\sigma_f} \cdots\rangle_{\text{rad}} \tag{6-41a}$$

$$|I_2\rangle = |\mathbf{q}_2, \lambda_2\rangle_e \, |\cdots n_{k_i\sigma_i}, \cdots n_{k_f\sigma_f} + 1 \cdots\rangle_{\text{rad}} \tag{6-41b}$$

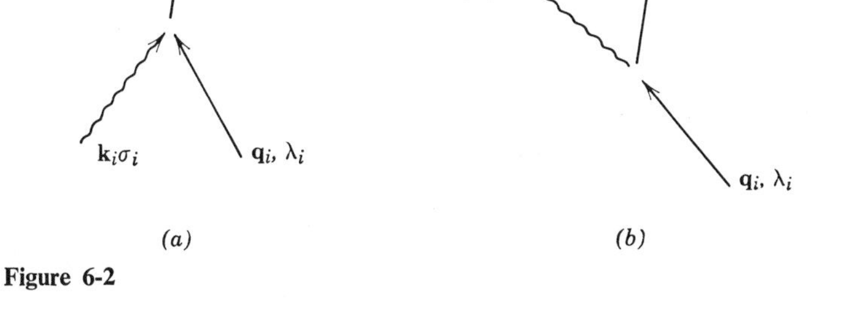

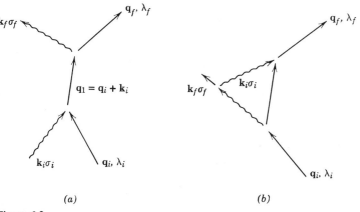

Figure 6-2

Conservation of momentum at the vertices gives

$$\mathbf{q}_1 = \mathbf{q}_i + \mathbf{k}_i = \mathbf{q}_f + \mathbf{k}_f \tag{6-42a}$$

$$\mathbf{q}_2 = \mathbf{q}_i - \mathbf{k}_f = \mathbf{q}_f - \mathbf{k}_i \tag{6-42b}$$

the transition probability per unit time for the process is given by

$$\left(\frac{\text{trans prob}}{\text{time}}\right) = \frac{2\pi}{\hbar} |M_{fi}|^2 \times \delta[\sqrt{\hbar^2c^2q_i^{\,2} + m^2c^4} - \sqrt{\hbar^2c^2q_f^{\,2} + m^2c^4} + \hbar ck_i - \hbar ck_f] \tag{6-43}$$

where M_{fi} is given by the second term in Eq. 1-166:

$$M_{fi} = \sum_{\lambda_1} \frac{\langle f|\, H_I\, |I_1\rangle\langle I_1|\, H_I\, |i\rangle}{E_i - E_{I_1}} + \sum_{\lambda_2} \frac{\langle f|\, H_I\, |I_2\rangle\langle I_2|\, H_I\, |i\rangle}{E_i - E_{I_2}} \tag{6-44}$$

the intermediate energies are

$$E_{I_1} = \pm\sqrt{\hbar^2c^2\, |\mathbf{q}_i + \mathbf{k}_i|^2 + m^2c^4} \tag{6-45a}$$

$$E_{I_2} = \pm\sqrt{\hbar^2c^2\, |\mathbf{q}_i - \mathbf{k}_f|^2 + m^2c^4} + \hbar ck_i + \hbar ck_f \tag{6-45b}$$

Using Eq. 6-14 we find that the necessary matrix elements are

$$\langle I_1|\, H_I\, |i\rangle = -e\left(\frac{2\pi\hbar c^2}{\Omega\omega_i}\right)^{1/2} (u^+_{\mathbf{q}_1\lambda_1}\boldsymbol{\alpha} \cdot \mathbf{u}_i u_{\mathbf{q}_i\lambda_i}) \tag{6-46a}$$

$$\langle f|\, H_I\, |I_1\rangle = -e\left(\frac{2\pi\hbar c^2}{\Omega\omega_f}\right)^{1/2} (u^+_{q_f\lambda_f}\boldsymbol{\alpha} \cdot \mathbf{u}_f u_{\mathbf{q}_1\lambda_1}) \tag{6-46b}$$

$$\langle I_2|\, H_I\, |i\rangle = -e\left(\frac{2\pi\hbar c^2}{\Omega\omega_f}\right)^{1/2} (u^+_{\mathbf{q}_2\lambda_2}\boldsymbol{\alpha} \cdot \mathbf{u}_f u_{q_1\lambda_1}) \tag{6-46c}$$

$$\langle f|\, H_I\, |I_2\rangle = -e\left(\frac{2\pi\hbar c^2}{\Omega\omega_i}\right)^{1/2} (u^+_{q_f\lambda_f}\boldsymbol{\alpha} \cdot \mathbf{u}_i u^+_{\mathbf{q}_2\lambda_2}) \tag{6-46d}$$

Equation 6-44 contains summations over λ_1 and λ_2, the spins and signs of the energies of the intermediate states.

Before proceeding to the general case let us consider the nonrelativistic limit. In this limit

$$u_{\mathbf{q},1} = \begin{pmatrix} 1 \\ 0 \\ 0 \\ 0 \end{pmatrix}, \qquad u_{\mathbf{q},2} = \begin{pmatrix} 0 \\ 1 \\ 0 \\ 0 \end{pmatrix} \tag{6-47a}$$

$$u_{\mathbf{q},3} = \begin{pmatrix} 0 \\ 0 \\ 1 \\ 0 \end{pmatrix}, \qquad u_{\mathbf{q},4} = \begin{pmatrix} 0 \\ 0 \\ 0 \\ 1 \end{pmatrix} \tag{6-47b}$$

Now, since

$$\boldsymbol{\alpha} \cdot \mathbf{u} = \begin{pmatrix} 0 & \boldsymbol{\sigma} \cdot \mathbf{u} \\ & 0 \\ \boldsymbol{\sigma} \cdot \mathbf{u} & \end{pmatrix} = \begin{bmatrix} 0 & 0 & u_z & u_- \\ 0 & 0 & u_+ & -u_z \\ u_z & u_- & 0 & 0 \\ u_+ & -u_z & 0 & 0 \end{bmatrix} \tag{6-48}$$

where $u_\pm = u_x \pm iu_y$, we see that the operator couples positive energy states to negative energy states only. The intermediate states must be states of negative energy for the scattering of photons by positive energy electrons.

Let us suppose that the initial and final states of the electron both have spin up. Then they are described by $u_{\mathbf{q},1}$ of Eq. 6-47a. In the sum over λ_1 in Eq. 6-44 only the terms with $\lambda_1 = 3, 4$ have nonvanishing matrix elements, and E_{I_1} is the same for both of these. A little algebra shows that

$$\sum_{\lambda_1=3,4} (u_{qf1}^{+}\boldsymbol{\alpha} \cdot \mathbf{u}_f u_{q_1\lambda_1})(u_{q_1\lambda_1}^{+}\boldsymbol{\alpha} \cdot \mathbf{u}_i u_{\mathbf{q}_i 1}) = \mathbf{u}_f \cdot \mathbf{u}_i \tag{6-49}$$

A similar result is obtained from the second term in Eq. 6-44. The matrix element becomes

$$M_{fi} = e^2\left(\frac{2\pi\hbar c^2}{\Omega}\right)\frac{\mathbf{u}_i \cdot \mathbf{u}_f}{\sqrt{\omega_i\omega_f}}$$
$$\times \left\{\frac{1}{\sqrt{\hbar^2c^2{q_i}^2 + m^2c^2} + \hbar ck_i + \sqrt{\hbar^2c^2\,|\mathbf{q}_i + \mathbf{k}_i|^2 + m^2c^4}}\right.$$
$$\left. + \frac{1}{\sqrt{\hbar^2c^2{q_i}^2 + m^2c^4} + \sqrt{\hbar^2c^2\,|\mathbf{q}_i - k_f|^2} - \hbar ck_i - \hbar ck_f}\right\} \tag{6-50}$$

In the nonrelativistic limit this becomes

$$M_{fi} = e^2\left(\frac{2\pi\hbar c^2}{\Omega}\right)\frac{\mathbf{u}_i \cdot \mathbf{u}_f}{\sqrt{\omega_i\omega_f}}\frac{1}{mc^2} \tag{6-51}$$

This is the same matrix element which appears in Eq. 3-39 of our previous calculation of scattering by nonrelativistic electrons, so that the two theories agree, as they should, in the nonrelativistic limit. It is noteworthy that negative energy states must occur as intermediate states in order to get this agreement. This shows that negative energy states are essential for the theory.

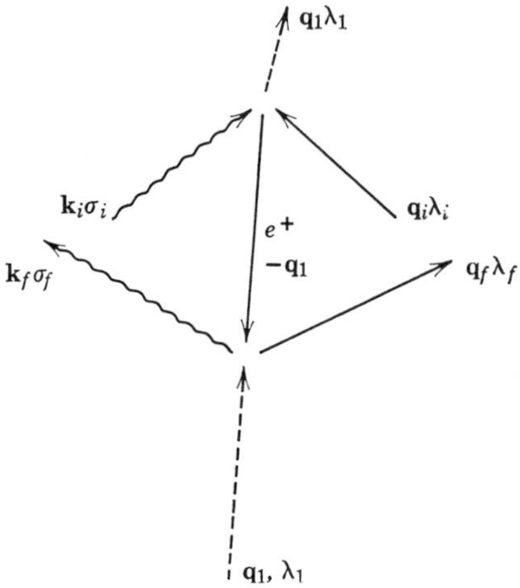

Figure 6-3

One feature of the calculation seems to be inconsistent with our previous assertions. We have allowed transitions to occur to negative energy states which we have assumed to be filled. It would appear that such transitions should be forbidden by the exclusion principle. The intermediate states with electrons of negative energy as we have drawn them in Fig. 6-2 cannot, in fact, exist. However, there are two other intermediate states that we have ignored which make the same contribution to M_{fi} as the two improper states which we have incorrectly included. To see this consider the diagram in Fig. 6-3.

In this process an unobservable negative energy electron of momentum $h\mathbf{q}_1$ emits the final photon at the first vertex becoming the final electron of momentum $h\mathbf{q}_f$. There is left behind a hole which to an observer would appear as a positron of momentum $-h\mathbf{q}_1$. We have drawn the unobservable negative energy electron as a dotted line and the observable hole as a solid line directed downward toward the first vertex. At the second vertex the primary electron of momentum $\hbar\mathbf{q}_i$ absorbs the primary photon and jumps into the hole thereby filling the negative energy state of momentum $\hbar\mathbf{q}_1$. The matrix elements are the same as those for Fig. 6-2*a*, since the vertices are the same, but they occur in the opposite order. The initial and final energies are the same, namely

$$E_i = E_{q_i} + \hbar\omega_i \tag{6-51a}$$

$$E_f = E_{q_f} + \hbar\omega_f \tag{6-51b}$$

The intermediate energy in Fig. 6-2*a* is

$$E_I = -|E_{q_1}| \tag{6-51c}$$

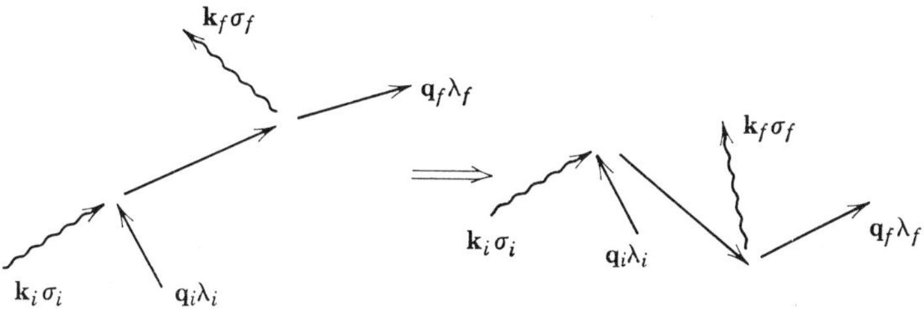

Figure 6-4

since the electron in the intermediate state is assumed to have negative energy. Then

$$E_i - E_I = E_{q_i} + \hbar\omega_i + |E_{q_1}| \tag{6-51d}$$

the intermediate state in Fig. 6-3 has the energy

$$E'_I = E_{qf} + |E_{q_1}| + \hbar\omega_i + \hbar\omega_f + E_{qi} \tag{6-52a}$$

since there is overall conservation of energy, $E_i = E_f$ and

$$E_i - E'_I = E_f - E'_I = -E_{qi} - \hbar\omega_i - |E_{qi}| \tag{6-52b}$$

which is just the negative of Eq. 6-51d. Now, the final state in Fig. 6-3 is not quite the same as the final state in Fig. 6-2*a*. In Fig. 6-3 there has been an exchange of the primary electron with one of the negative energy electrons. The exchange of fermions gives a factor of -1 which just compensates for the change in sign of the energy denominator. (This is one of the cases when the phase factors in Eq. 4-24 must be taken into account.) The result in that the allowed process of Fig. 6-3 makes the same contribution to M_{fi} as the not allowed process of Fig. 6-2*a*. Note that when the dotted lines are omitted Fig. 6-3 can be obtained from Fig. 6-2*a* by rotating the arrow representing the intermediate state around until it points in the negative time direction. This is illustrated in Fig. 6-4. A positron may be thought of as an electron propagating backward in time.

Problem 6-2. Draw the diagram which is equivalent to Fig. 6-2*b* in the same sense that Fig. 6-3 is equivalent to Fig. 6-2*a*.

It seems to be usually true, that we obtain the same results by allowing negative energy particles as intermediate states, as we do by using hole theory, but this should be checked in each case.

The matrix element M_{fi} can be evaluated without making the nonrelativistic approximation. Its square is then used in the Fermi golden rule and the scattering cross section is obtained. These calculations involve some of the tricks developed in the preceding section. They are fairly tedious and have been relegated to the appendix. The result is the well known Klein-Nishina[34] formula

$$\frac{d\sigma}{d\Omega} = \left(\frac{e^2}{mc^2}\right)^2 \frac{k_f^{\,2}}{k_i^{\,2}}\left[(\mathbf{u}_i \cdot \mathbf{u}_f)^2 - \frac{1}{2} + \frac{1}{4}\frac{k_i}{k_f} + \frac{1}{4}\frac{k_f}{k_i}\right] \tag{6-53}$$

the relation between the initial and final wave numbers is

$$\frac{1}{k_f} - \frac{1}{k_i} = \frac{\hbar}{mc}(1 - \cos\theta) \tag{6-54}$$

in the classical limit ($\hbar \rightarrow 0$) and in the long wavelength limit, $k_i \simeq k_f$, the last three terms in Eq. 6-53 cancel and the equation agrees with the Thompson scattering cross section of Eq. 3-46.

PAIR PRODUCTION

It is not possible for a photon in free space to create an electron-positron pair because in doing so it would violate the conservation laws for momentum and energy. However, in the presence of a third body which can carry off some momentum this pair production process can occur. It is analogous to the inverse of the bremsstrahlung process discussed in Chapter 5. We may picture the process as occurring as shown in the Feynman diagram of Fig. 6-5. In this process one of the unobservable negative energy electrons represented by a dotted line collides with the third body, which we assume to be a heavy nucleus representable by a potential V. In this collision the electrons momentum is changed from $h\mathbf{q}_1$, to $h\mathbf{q}_2$ and λ changes from 3 or 4 to 1 or 2. Since its energy is positive after the collision, the electron is observable so we represent it by a solid line. Also the hole left behind is observable as a positron, so we have denoted it by a solid line directed toward the first vertex. At the second vertex the electron absorbs the photon changing its momentum from $\hbar\mathbf{q}_2$ to $\hbar\mathbf{q}_3 = \hbar(\mathbf{q}_2 + \mathbf{k})$. The net result is that a photon is absorbed and an electron-positron pair is created. Note that when the dotted line is omitted

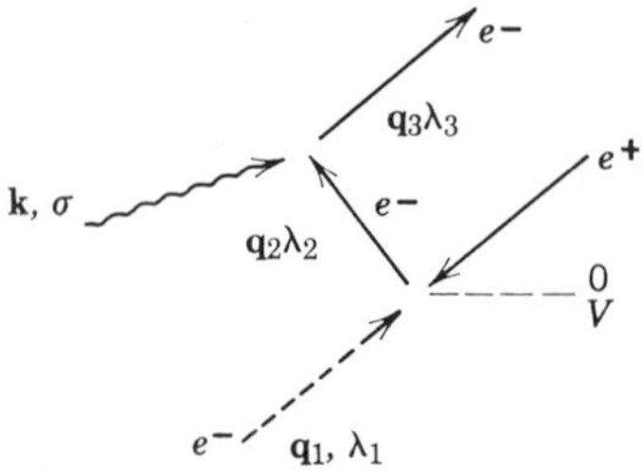

Figure 6-5

the diagram looks just like that for inverse bremstrahlung except that positron line is directed downward in time.

The calculation of the cross section for this process is very similar to that for bremsstrahlung treated in Chapter 5. Of course, one must use Eq. 6-14 in calculating the photon absorption. Equation 5-13 must involve the Dirac spinors. The calculation is somewhat more tedious than that of relativistic bremsstrahlung and we shall not give the details.

ELECTRON-POSITRON ANNIHILATION

Electrons and positrons can annihilate by the inverse of the process shown in Fig. 6-5. An electron emits a photon and jumps into the empty negative energy state. The presence of a third body is necessary to conserve energy and momentum. This process sometimes occurs when a positron collides with an electron bound in an atom.

Free electrons and positrons can also annihilate by a second order process in which two photons are emitted. The process may be pictured as occurring as shown in the Feynman diagram of Fig. 6-6. Initially, there is present an electron of momentum $h\mathbf{q}_1$ and a positron of momentum $-h\mathbf{q}_2$. This really indicates that a negative energy electron of momentum $h\mathbf{q}_2$ is missing from the negative energy sea. At the first vertex the electron emits a photon. At the second vertex it emits a second photon and jumps into the negative energy state of momentum $h\mathbf{q}_2$ which was previously vacant. As usual we have shown observable particles as solid lines and unobservable particles as dotted lines. Note that the diagram looks very much like the diagram for Compton scattering except that the positron line is directed in the negative time direction.

The calculation proceeds very much as the Compton scattering calculation. We shall simplify things a little by working in the center of mass system where $\mathbf{q}_2 = \mathbf{q}_1$. Then from momentum conservation $\mathbf{k}_2 = -\mathbf{k}_1$.

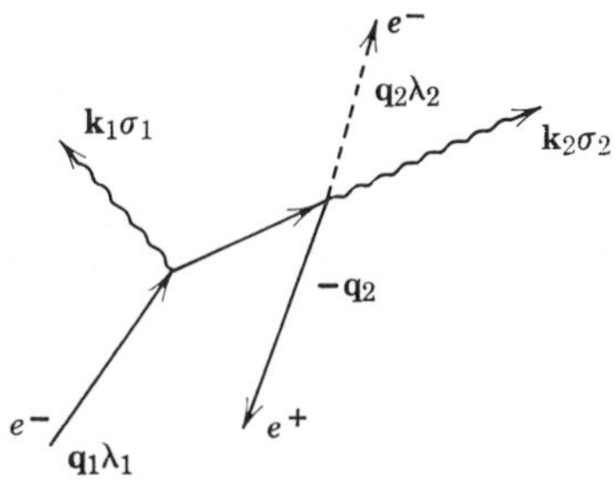

Figure 6-6

Conservation of energy gives

$$E_i = E_f \tag{6-55a}$$

$$2\sqrt{\hbar^2c^2q_1^{\,2} + m^2c^4} = 2\hbar c k_1 \tag{6-55b}$$

from which

$$k_1 = \sqrt{q_1^{\,2} + (mc/\hbar)^2} \tag{6-56}$$

We suppose that the electron and positron are moving very slowly so that $q_1 \simeq 0$. We can write the lifetime of the positron as

$$\begin{aligned} \frac{1}{\tau} &= \sum_{\text{final states}} \frac{2\pi}{\hbar} |M|^2 \, \delta(E_i - E_f) \\ &= \frac{\Omega}{(2\pi)^3} \frac{2\pi}{\hbar} \sum \int k_1^{\,2} \, dk_1 \, d\Omega_k \, |M|^2 \frac{1}{2\hbar c} \, \delta\left(k_1 - \sqrt{q_1^{\,2} + \left(\frac{mc}{\hbar}\right)^2}\right) \\ &= \frac{\Omega}{(2\pi)^3} \frac{2\pi}{\hbar} \frac{1}{2\hbar c}\left[q_1^{\,2} + \left(\frac{mc}{h}\right)^2\right] \sum \int d\Omega_k \, |M|^2 \end{aligned} \tag{6-57}$$

In this formula M is the matrix element for the transition and $d\Omega_k$ is the solid angle into which the photon of momentum $\hbar k_1$ is emitted. The other photon goes in the opposite direction. The summation in the final formula is over polarizations. We have used Eq. 3-12 to replace a sum by an integral. The matrix element may be written

$$M = e^2\left(\frac{2\pi\hbar c^2}{\Omega\omega}\right) \sum_I \frac{\langle f|\, H_I' \,|I\rangle \langle I|\, H_I' \,|i\rangle}{E_i - E_I} \tag{6-58}$$

where we have denoted the part of Eq. 6-14 without the factor $e(2\pi\hbar c^2/\Omega\omega)^{1/2}$ by H_I'. The sum is over the quantum numbers λ which the electron can have in the intermediate state and also the other time order in which the electron can emit the photons.

Since $q_1 \simeq 0$, the momentum of the electron in the intermediate state must be $\hbar k_1 = mc$; this makes its energy $\sqrt{2}mc^2$. Then

$$E_i = 2mc^2 \tag{6-59a}$$

$$E_I = \sqrt{2}mc^2 + mc^2 + \hbar c k_1 = \sqrt{2}mc^2 + 2mc^2 \tag{6-59b}$$

$$E_i - E_I = \sqrt{2}mc^2 \tag{6-59c}$$

The matrix elements in Eq. 6-58 involve only the Dirac matrices and spinors, so they must be of order unity. Therefore, as an order of magnitude estimate of M we may take

$$M \simeq e^2\left(\frac{2\pi\hbar c^2}{\Omega\omega}\right)^2 \frac{1}{mc^2} \tag{6-60}$$

using this in Eq. 6-57 and replacing the integration over solid angle by 4π we obtain the order of magnitude estimate

$$\frac{1}{\tau} = \frac{\pi c}{\Omega}\left(\frac{e^2}{mc^2}\right)^2 = \frac{\pi c}{\Omega} r_e^{\,2} \tag{6-61}$$

This result agrees with an exact calculation.

It may seem strange at first sight to find Ω, the volume of the box in which the system was quantized in the final formula, but it must be remembered that there is only one electron in the box. The electron density is then $n = 1/\Omega$, and the formula should read

$$\frac{1}{\tau} = \pi r_e^{\,2} n c \tag{6-62}$$

Taking $n \simeq 10^{24}\ \text{cm}^{-3}$ and $r_e \simeq 10^{-13}$ cm gives $\tau \simeq 10^{-9}$ sec as the lifetime of a positron in a solid.

A positron and electron can also form a bound state of the hydrogen-like atom positronium before it decays. In the first approximation the levels and wave functions of positronium are those of a hydrogen atom with the Bohr radius replaced by $a = 2a_0 = 2\hbar^2/me^2$ because of the reduced mass. Equation 6-62 can be used to estimate the lifetime of a positronium atom if we take

$$n = |\psi_{1s}(0)|^2 = \frac{1}{\pi a^3} \tag{6-63}$$

this gives $\tau \simeq 10^{-10}$ sec. However, the positronium atom can only annihilate by two-photon emission when the spins are antiparallel (the ^{1}S state). If the atom is in the ^{3}S state it must annihilate by three-photon emission. The lifetime for this mode of annihilation is longer by about a factor of 370.

7

The Theory of Beta Decay

The success of quantum field theory in describing processes in which photons, electrons, and positrons are created and destroyed suggest that the theory could be extended to similar processes in nuclear and high energy physics. There are two difficulties in making such an extension. First, we do not have a classical theory to guide us, as we did in the case of the electromagnetic field. In order for a field to behave clasically, it must be possible to put a large number of quanta in the same state. Consequently, the fields describing fermions can never behave classically. In principle the field that describes mesons could have a classical limit, but because of the short lifetime of these particles it is impractical to put a large number of mesons in the same state.

The second difficulty has to do with the strength of the interaction. The interaction of photons and electrons is in a sense weak, and perturbation theory gives excellent results. In quantum electrodynamics, perturbation theory may be regarded as an expansion in the fine structure constant $e^2/\hbar c = \frac{1}{137}$ which is a small number. The corresponding expansion parameter for the so-called strong interaction responsible for nuclear forces is much larger than this; as a result, perturbation theory is almost useless. The weak interaction which is responsible for beta decay is characterized by an expansion parameter which is much smaller than $e^2/\hbar c$.

Consequently, perturbation theory is applicable to the weak interaction. The theory of beta decay which has been developed in analogy with quantum electrodynamics is remarkably successful. It is the subject of this chapter.

The first theory of beta decay was proposed by Enrico Fermi in 1933. At this time a very puzzling feature of beta decay was well known and was the subject of much discussion. This was the fact that the electron emitted in the decay emerged with a continuous energy spectrum. One would expect that the electron would have an energy equal to the difference of the energies of the parent and daughter nuclei. Instead, it was found that the maximum energy of the electron was equal to this difference, but that the energy spectrum extended continuously down to zero energy. Pauli had made (but not published) the suggestion that this could be accounted for if another particle (which is now called the neutrino) was emitted at the same time as the electron. The available energy would be shared between the electron and neutrino. To conserve charge the neutrino must be assumed to be neutral. This would make its detection difficult and explain why it had escaped detection. Indeed, it continued to escape detection for another twenty years.

It was also difficult to explain how electrons could be bound in orbits of nuclear dimensions. The kinetic energies of electrons in such small orbits would be much greater than the observed energies of escaping electrons. Today there is even greater evidence based on spin and statistics that electrons cannot be present in the nucleus.

Fermi disposed of these difficulties by assuming that in the decay process an electron and a neutrino (actually, it is preferable to regard it as an antineutrino) are created, as a neutron changes to a proton. This basic process can be represented by the diagram of Fig. 7-1. We have denoted the neutron by n. proton by p, electron by e, and antineutrino by $\bar{\nu}$. To produce such a process the interaction Hamiltonian must be somewhat like

$$H_I = g \int d^3x \psi_p^+ \psi_n \psi_e^+ \psi_{\bar{\nu}}^+ + \text{HC} \tag{7-1}$$

That is, the integrand must contain as factors operators which destroy a neutron and create a proton, an electron, and an antineutrino. The constant g is a measure of the strength of the interaction. The Hermitian conjugate term (HC) would destroy a proton, an electron, and an antineutrino and create a neutron. If the electron which was destroyed was an unobservable negative energy electron, then this would be equivalent to creation of a positron, and the HC term could describe positron emission.

All four of the particles involved in this process are Fermions, so in a correct relativistic theory ψ_p, ψ_n, ψ_e, and ψ_ν should all be Dirac spinors. In choosing the way in which they should be combined, Fermi was guided by considerations of relativistic invariance and analogy with quantum electrodynamics. In Eq. 6-18 we wrote the integrand of the electron-photon interaction Hamiltonian as the scalar product of two 4-vectors. One of these was

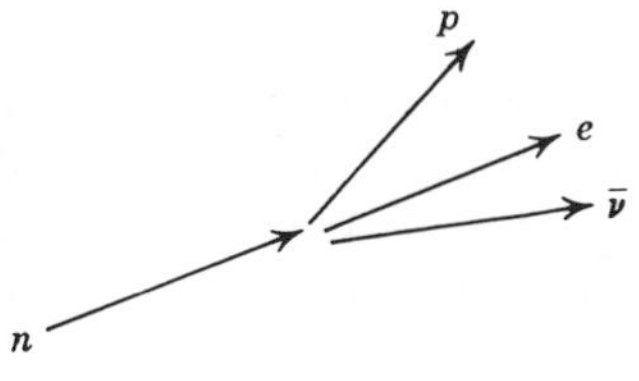

Figure 7-1

the potential A_μ of the photon field; the other was the electron current vector $-e_p\bar{\psi}\gamma_\mu\psi$. We may call this a vector interaction. In constructing an interaction Hamiltonian for beta decay we have four Fermion operators available to us. These may be used in pairs to construct 4-vectors; then the scalar product of these four vectors is the integrand of H_I. In this way we obtain

$$H_I = g\int d^3x(\bar{\psi}_p\gamma_\mu\psi_n)(\bar{\psi}_e\gamma_\mu\psi_\nu) + \text{HC} \tag{7-2}$$

(The operator ψ_ν in Eq. 7-2 can destroy a negative energy neutrino; this is equivalent to creating an antineutrino.) This is not the only plausible form for H_I; we discuss other forms and what is believed to be the correct form later in this chapter.

For a first calculation we use Eq. 7-1. This is useful in showing how much can be accomplished with very little theory. We expand ψ_p^+ and ψ_n in the states ϕ_{pa}^* and ϕ_{pb} which may be bound states of the proton and neutron in a nucleus. Thus

$$\psi_p^+(x) = \sum_a A_{pa}^+\phi_{pa}^*(\mathbf{x}) \tag{7-3a}$$

$$\psi_n(x) = \sum_b A_{nb}\phi_{nb}(\mathbf{x}) \tag{7-3b}$$

Where A_{pa}^+ is a creation operation for a proton in state $|a\rangle_p$, and A_{nb} is an annihilation operator for a neutron in state $|b\rangle_n$. We shall expand ψ_e^+ and $\psi_{\bar{\nu}}^+$ in plane wave functions; thus

$$\psi_e^+(\mathbf{x}) = \sum_{\mathbf{q}_e} A_{\mathbf{q}_e}^+ \frac{e^{i\mathbf{q}_e\cdot\mathbf{x}}}{\sqrt{\Omega}} \tag{7-4a}$$

$$\psi_{\bar{\nu}}^+(\mathbf{x}) = \sum_{\mathbf{q}_\nu} A_{\mathbf{q}_{\bar{\nu}}}^+ \frac{e^{i\mathbf{q}_{\bar{\nu}}\cdot\mathbf{x}}}{\sqrt{\Omega}} \tag{7-4b}$$

Where $A_{\mathbf{q}_e}^+$ and $A_{\mathbf{q}_{\bar{\nu}}}^+$ are creation operators for electrons and antineutrinos of momentum $\hbar\mathbf{q}_e$ and $\hbar\mathbf{q}_{\bar{\nu}}$ respectively. Equation 7-1 becomes

$$H_I = \frac{g}{\Omega}\sum_a\sum_b\sum_{\mathbf{q}_e}\sum_{\mathbf{q}_{\bar{\nu}}} M_{ab}A_{pa}^+A_{nb}A_{\mathbf{q}_e}^+A_{\mathbf{q}_{\bar{\nu}}}^+ + \text{HC} \tag{7-5}$$

where

$$M_{ab} = \int d^3x\phi_{pa}^*\phi_{nb}e^{-i(\mathbf{q}_e+\mathbf{q}_\nu)\cdot\mathbf{x}} \tag{7-6}$$

In calculating the transition probabilities per unit time we use the relativistic expressions for the energies of the electron and the neutrino. We assume that the mass of the neutrino is zero, so that its energy is $\hbar c q_{\bar{\nu}}$. To get the lifetime for the decay we sum over the final states of the electron and antineutrino and use Eq. 3-12 to convert the sums to integrals. We obtain

$$\begin{aligned}\frac{1}{\tau} &= \sum_{\mathbf{q}_e}\sum_{\mathbf{q}_\nu}\left(\frac{\text{trans Prob}}{\text{time}}\right)\\ &= \frac{\Omega^2}{(2\pi)^6}\frac{2\pi}{\hbar}\frac{|g|^2}{\Omega^2}\int d^3q_e\int d^3q_{\bar{\nu}}\\ &\quad\times |M_{ab}|^2\,\delta[E_{nb}-E_{pa}-\sqrt{\hbar^2c^2q_e^{\,2}+m^2c^4}-\hbar c q_{\bar{\nu}}]\end{aligned} \tag{7-7}$$

Now, let us examine Eq. 7-6. The wavelengths of electrons and neutrinos emitted in beta decays are usually much larger than nuclear dimensions. Therefore, it is usually a good approximation to replace the exponential in the integrand by unity; thus

$$M_{ab} \simeq \int d^3x\,\phi_{pa}^*\phi_{nb} \tag{7-8}$$

This is independent of $\mathbf{q}_e$ and $q_{\bar{\nu}}$, so that it can be removed from the integral. The integration over solid angles and over $q_{\bar{\nu}}$ can be carried out and there remains

$$\frac{1}{\tau} = \frac{|g|^2\,|M_{ab}|^2}{2\pi^3\hbar^4c^3}\int dq_e q_e^{\,2}[E_{\max}-E_e]^2 \tag{7-9}$$

where $E_{\max} = E_{nb} - E_{na}$ is maximum energy the electron can have, and $E_e = \sqrt{\hbar^2c^2q_e^2 + m^2c^4}$. We can write this as

$$\frac{1}{\tau} = \int dp_e I(p_e) \tag{7-10a}$$

where

$$I(p_e)\,dp_e = \frac{|g|^2\,|M_{ab}|^2}{2\pi^3\hbar^7c^3}\,p_e^{\,2}[E_{\max}-E_e]^2 \tag{7-10b}$$

is the probability of decay with emission of an electron with momentum between p_e and $p_e + dp_e$.

The theory may be checked by plotting $[I(p_e)/p_e^{\,2}]^{1/2}$ versus E_e. The result should be a straight line which intersects the axis at $E = E_{\max}$. This is called a Kurie plot. Also, the integration in Eq. 7-9 can be carried out. The result is a function of $E_{\max}$. Actually, the theory given here is cruder than it needs to be. One should use the Coulomb wave functions for the electron rather than the free particle wave functions. The integral will then be a function of the atomic number Z as well as of $E_{\max}$. We denote this by $f(Z, E_{\max})$. Then

$$\frac{1}{\tau f(Z, E_{\max})} = \frac{|g|^2\,|M_{ab}|^2}{2\pi^3\hbar^7c^3} \tag{7-11}$$

Thus the product of $\tau f(Z, E_{\max})$ may be used as a measure of $|g|\,|M_{ab}|^2$.

Now, let us return to the Fermi theory with H_I given by Eq. 7-2. Recalling that $\gamma_\mu = (-i\beta\boldsymbol{\alpha}, \beta)$ and $\bar{\psi} = i\psi^+\beta$ we can write

$$(\bar{\psi}_p\gamma_\mu\psi_n)(\bar{\psi}_e\gamma_\mu\psi_\nu) = (\psi_p^+\boldsymbol{\alpha}\psi_n)\cdot(\psi_e^+\boldsymbol{\alpha}\psi_\nu) - (\psi_p^+\psi_n)(\psi_e^+\psi_\nu) \qquad (7\text{-}12)$$

It is a good approximation to treat the nucleons nonrelativistically so that

$$\psi_n \sim \begin{bmatrix}1\\0\\0\\0\end{bmatrix} \quad \text{or} \quad \begin{bmatrix}0\\1\\0\\0\end{bmatrix} \qquad (7\text{-}13a)$$

and

$$\psi_p^+ \sim [1, 0, 0\ \ 0] \quad \text{or} \quad [0, 1, 0, 0] \qquad (7\text{-}13b)$$

Since

$$\boldsymbol{\alpha} = \begin{bmatrix}0 & \boldsymbol{\sigma}\\ \boldsymbol{\sigma} & 0\end{bmatrix} \qquad (7\text{-}14)$$

it follows that

$$\psi_p^+\boldsymbol{\alpha}\psi_n = 0 \qquad (7\text{-}15)$$

Then in the approximation that the nucleons can be treated nonrelativistically, the interaction Hamiltonian simplifies to

$$H_I = -g\int d^3x\psi_p^+\psi_n\psi_e^+\psi_\nu \qquad (7\text{-}16)$$

We also see from Eq. 7-13 that $\psi_p^+\psi_n$ will vanish unless the spins of the proton and neutron are the same.

Now, we expand just as we did in Eqs. 7-3 and 7-4 but use the Dirac wave functions for electrons and neutrinos. The matrix element for the transition is

$$\langle f|\, H_I\, |i\rangle = \frac{g}{\Omega} M_{ab} u^+_{\mathbf{p}_e\lambda_e} u_{\mathbf{p}_\nu\lambda_\nu} \qquad (7\text{-}17)$$

It differs from the previous result by the product of the Dirac spinors for the electron and the neutrino.

Since the spins of the electrons and neutrinos emitted in beta decays are usually not measured, we would like to sum the square of Eq. 7-17 over the spins of the electron and the neutrino. Thus the appropriate square matrix element is

$$|M|^2 = \frac{|g|^2}{\Omega^2}|M_{ab}|^2 \sum_{\lambda_e=1,2}\ \sum_{\lambda_\nu=3,4} u^+_{\mathbf{p}_e\lambda_e} u_{\mathbf{p}_\nu\lambda_\nu} u^+_{\mathbf{p}_\nu\lambda_\nu} u_{\mathbf{p}_e\lambda_e} \qquad (7\text{-}18)$$

In evaluating this we can use some of the tricks developed in the preceding chapter. We use

$$\left(\frac{H + |E|}{2\,|E|}\right) u_{\mathbf{p}\lambda} = \begin{cases} u_{\mathbf{p}\lambda} & \lambda = 1, 2 \\ 0 & \lambda = 3, 4 \end{cases} \tag{7-19a}$$

and

$$\left(\frac{H - |E|}{-2\,|E|}\right) u_{\mathbf{p}\lambda} = \begin{cases} 0 & \lambda = 1, 2 \\ u_{\mathbf{p}\lambda} & \lambda = 3, 4 \end{cases} \tag{7-19b}$$

To write

$$|M|^2 = \frac{-|g|^2\,|M_{ab}|^2}{\Omega^2 4\,|E_{p_e}|\,|E_{p_\nu}|} \sum_{\lambda_e=1}^{4} \sum_{\lambda_\nu=1}^{4} \{u^{+}_{\mathbf{p}_e\lambda_e}(H_{\mathbf{p}_\nu} - |E_{\mathbf{p}_\nu}|)u_{p_\nu\lambda_\nu}u^{+}_{\mathbf{p}_\nu\lambda_\nu}(H_{\mathbf{p}_e} + |E_{\mathbf{p}_e}|)u_{\mathbf{p}_e\lambda_e}\} \tag{7-20}$$

Now using

$$\sum_{\lambda_\nu=1}^{4} u_{\mathbf{p}_\nu\lambda_\nu}u^{+}_{\mathbf{p}_\nu\lambda_\nu} = 1 \tag{7-21}$$

we can reduce Eq. 7-20 to

$$|M|^2 = \frac{|g|^2\,|M_{ab}|^2}{\Omega^2 4\,|E_e|\,|E_\nu|} \operatorname{Tr}(|E_\nu| - H_\nu)(|E_e| - H_e) \tag{7-22}$$

where we have dropped some superfluous subscripts. Now

$$\operatorname{Tr} H_\nu = \operatorname{Tr} H_e = 0 \tag{7-23}$$

Since H_ν and H_e contain $\boldsymbol{\alpha}$ and β linearly, and the $\boldsymbol{\alpha}$ and β matrices have zero trace

$$\operatorname{Tr} |E_e|\,|E_\nu| = 4\,|E_e|\,|E_\nu| \tag{7-24}$$

and

$$\operatorname{Tr} H_\nu H_e = c^2 \operatorname{Tr}(\boldsymbol{\alpha}\cdot\mathbf{p}_\nu)(\boldsymbol{\alpha}\cdot\mathbf{p}_e) = 4c^2\mathbf{p}_e\cdot\mathbf{p}_\nu \tag{7-25}$$

where Eq. 6-23 has been used. Then, using $\mathbf{v} = c^2\mathbf{p}/|E|$ gives

$$|M|^2 = \frac{|g|^2\,|M_{ab}|^2}{\Omega^2}\left(1 - \frac{\mathbf{v}_e\cdot\mathbf{v}_\nu}{c^2}\right) \tag{7-26}$$

This is the appropriate square matrix element to use in calculating the lifetime. It differs from the previous result only by the factor $(1 - \mathbf{v}_e\cdot\mathbf{v}_\nu/c^2)$. When used in Eq. 7-7 the term in $\mathbf{v}_e\cdot\mathbf{v}_\nu$ vanishes when the neutrino momentum is integrated over, so that our previous result for the lifetime is unchanged.

Experimentally, it is found that the values of values of $\tau f(Z, E_{\max})$ fall in groups that are separated by one and more orders of magnitude. The decays with the smallest τf values have $\Delta I = 0$ (no change in the nuclear spin I). This is in agreement with the Fermi form of the interaction, since, as we have seen, this interaction does not change the spin of the nucleon. Also, M_{ab} as given by Eq. 7-8 will vanish if ϕ_{pa} and ϕ_{nb} have different orbital angular momenta. These decays with $\Delta I = 0$ have energy spectra which fit the Kurie plots very well. There are other decays which have $\Delta I \neq 0$ and τf values which are greater by several orders of magnitude. The energy spectra of these decays do not fit Kurie plots. These may be explained on the basis of the Fermi interaction by considering the terms which were neglected when Eq. 7-6 was approximated by Eq. 7-8. Expanding the exponential in Eq. 7-6, we may write

$$M_{ab} = \int d^3x \phi_{pa}^+ \phi_{nb}[|1 - i(\mathbf{q}_e + \mathbf{q}_\nu) \cdot \mathbf{x} - ((\mathbf{q}_e + \mathbf{q}_\nu) \cdot \mathbf{x})^2 + \cdots] \quad (7\text{-}27)$$

The first term couples only states of the same angular momentum; the second term couples states that differ by one unit of angular momentum; the third term couples states that differ by two units of angular momentum; and so on. These higher order terms account for the decays with $\Delta I = 1, 2$, and so on. The higher order terms should give smaller values of M_{ab}, hence larger values of τf. Also, the terms with $\Delta I \neq 0$ should have different energy spectra because of the additional factors of $(\mathbf{q}_e + \mathbf{q}_\nu)$ in M_{ab}. Therefore, it is not surprising that the Kurie plot is not fitted by the spectra of these decays.

There are exceptions to the scheme that we have just described. For example, the decay

$$_2^6\text{He} \rightarrow {}_3^6\text{Li} + e^- + \bar{\nu} \quad (7\text{-}28)$$

has $\Delta I = 1$ but about the same τf value as those decays for which $\Delta I = 0$. This suggests that the Fermi interaction is not completely correct and there are other terms in the interaction that change the spin of the nucleon. Only about two years after the publication of Fermi's paper, Gamow and Teller suggested other forms of the interaction which permit spin changes.

There are other relativistically invariant combinations of $\bar{\psi}_p$, ψ_n, $\bar{\psi}_e$, and ψ_ν than those of Eq. 7-2. We can form the scalar $\bar{\psi}_p\psi_n$ and multiply it by the scalar $\bar{\psi}_e\psi_\nu$ to get the scalar interaction

$$H_I^s = \int d^3x (\bar{\psi}_p\psi_n)(\bar{\psi}_e\psi_\nu) \quad (7\text{-}29a)$$

Or we can multiply the pseudoscalar $\bar{\psi}_p\gamma_5\psi_n$ by the pseudoscalar $\bar{\psi}_e\gamma_5\psi_\nu$ to get the pseudoscalar interaction

$$H_I^P = \int d^3x (\bar{\psi}_p\gamma_5\psi_\nu) \quad (7\text{-}29b)$$

The vector interaction

$$H_I^V = \int d^3x (\psi_p\gamma_\mu\psi_n)(\bar{\psi}_e\gamma_\mu\psi_\nu) \quad (7\text{-}29c)$$

is the one proposed by Fermi and has already been discussed. There is also the axial vector interaction

$$H_I^A = \int d^3x(\bar{\psi}_p\gamma_\mu\gamma_5\psi_n)(\bar{\psi}_e\gamma_\mu\gamma_5\psi_\nu) \tag{7-29d}$$

and the tensor interaction

$$H_I^T = \int d^3x(\bar{\psi}_p\sigma_{\mu\nu}\psi_n)(\bar{\psi}_e\sigma_{\mu\nu}\psi_\nu) \tag{7-29e}$$

where

$$\sigma_{\mu\nu} = \frac{i}{2}(\gamma_\mu\gamma_\nu - \gamma_\nu\gamma_\mu) \tag{7-30}$$

In constructing these forms of H the aim has been to make the integrand a scalar using only $\bar{\psi}_p$, ψ_n, $\bar{\psi}_e$, and ψ_ν and the γ matrices. This exhausts the possible scalar combinations (unless one introduces gradients, for instance) and at one time it was thought that H_I probably had the form

$$H_I = C_S H_I^S + C_P H_I^P + C_V H_I^V + C_A H_I^A + C_T H_I^T \tag{7-31}$$

The five coefficients $C_S \cdots C_T$ would have to be determined by experiment.

The problem of experimentally determining the five coefficients in Eq. 7-31 may seem bad enough, but then in 1956 Yang and Lee examined the evidence for the widespread assumption that parity was conserved in all interactions and concluded that in the case of the weak interactions there was no convincing experimental evidence for such an assumption. They suggested experiments to test parity conservation. These experiments were done and it was found that indeed parity was not conserved in beta decay. This indicated that there must be terms in H_I which behaved as pseudoscalars (i.e., changed sign under inversion) rather than scalars. For instance $(\bar{\psi}_p\gamma_5\psi_n)(\bar{\psi}_e\psi_\nu)$ is invariant under rotation and Lorentz transformation but changes sign under inversion. A term

$$H_i^{S'} = \int d^3x(\bar{\psi}_p\gamma_5\psi_n)(\bar{\psi}_e\psi_\nu) \tag{7-32}$$

would predict beta decays which violated parity conservation. The same applies for the other four couplings in Eq. 7-29. These should all be added to Eq. 7-31 with coefficients C_S', C_P', C_V', C_A', and C_T'. This gives ten coefficients to be determined by experiments.

We shall not recount the experimental and theoretical struggle which led to what is now believed to be the correct form of the beta interaction but shall just quote the result. The correct form is now believed to be

$$H_I = g\int d^3x(\psi_p\gamma_\mu a\psi_n)(\bar{\psi}_e\gamma_\mu a\psi_\nu) \tag{7-33a}$$

where

$$a = \tfrac{1}{2}(1 + i\gamma_5) \tag{7-33b}$$

and

$$g = 6.2 \times 10^{-44} \text{ MeV cm}^3 \tag{7-33c}$$

This differs from the Fermi form of the interaction only in having a included in each factor. This form of H_I has been very successful in accurately describing beta decays over an enormous range of lifetimes and decay energies. The term $(\bar{\psi}_p\gamma_\mu\gamma_5\psi_n)$ gives the Gamow-Teller part of the interaction in which the spin of the nucleon is changed.

Problem 7-1. Show that in the nonrelativistic approximation

$$\bar{\psi}_p i\gamma_\mu\gamma_5\psi_n = \begin{cases} \bar{\psi}_p\sigma_\mu\psi_n & m_\mu = 1, 2, 3 \\ 0 & \mu = 4 \end{cases} \tag{7-34}$$

It is now believed that there is a universal weak interaction of the form

$$H_i = g\int d^3x J_\mu J_\mu \tag{7-35a}$$

with

$$J_\mu = (\bar{\psi}_p\gamma_\mu a\psi_n) + (\bar{\psi}_e\gamma_\mu a\psi_\nu) + (\text{other terms}) \tag{7-35b}$$

The other terms in the current include terms responsible for the decay of muons and strange particles. It is the cross product term between $(\bar{\psi}_p\gamma_\mu a\psi_n)$ and $(\bar{\psi}_e\gamma_\mu a\psi_\nu)$ that causes the nuclear beta decays which we have just discussed.

Problem 7-2. The Σ° hyperon decays to the Λ hyperon with the emission of a γ-ray. Since both particles are uncharged, presumably the interaction with the electromagnetic field is through a magnetic moment. A reasonable guess as to the interaction Hamiltonian is

$$H_I = g\frac{e\hbar}{2Mc}\tau\boldsymbol{\sigma}\cdot(\boldsymbol{\nabla}\times\mathbf{A})$$

where M is of the order of the mass of the Σ° or the Λ, g is of order of unity, and τ is an operator that converts the Σ° into a Λ leaving the spin unchanged. Estimate the mean lifetime in seconds.

8

Particles that Interact Among Themselves

In Chapter 4 we developed a formalism for describing an arbitrary number of noninteracting nonrelativistic particles moving in an external potential $V(\mathbf{x})$. We can modify this formalism so as to take into account two body interactions among the particles by a simple addition of a term to the Hamiltonian of Eq. 4-5. The new Hamiltonian is

$$H = \int d^3x \psi^+(\mathbf{x}, t)\left[-\frac{\hbar^2}{2m}\nabla^2 + V(\mathbf{x})\right]\psi(\mathbf{x}, t) + \int d^3x \int d^3x' \psi^+(\mathbf{x}, t)\psi^+(\mathbf{x}, t)v(\mathbf{x}, \mathbf{x}')\psi(\mathbf{x}, t)\psi(\mathbf{x}', t) \quad (8\text{-}1)$$

where $v(\mathbf{x}, \mathbf{x}')$ is the potential energy of interaction between a particle at $\mathbf{x}$ and another at $\mathbf{x}'$. If we define an n-particle state vector by Eq. 4-43 and require that it be an eigenvector of H, then by steps similar to those that led to Eq. 4-45 we find that $C_n(x_1 \cdots x_n)$ must be a solution of

$$\left[-\sum_i^n \frac{\hbar^2}{2m}\nabla_i^2 + \sum_i^n V(\mathbf{x}_i) + \tfrac{1}{2}\sum_{i+j}^n\sum^n v(\mathbf{x}_i\mathbf{x}_j)\right]C_n = EC_n \quad (8\text{-}2)$$

which is the time independent Schrödinger equation for n interacting particles.

Now let us suppose that $V(\mathbf{x}) = 0$ and $v(\mathbf{x}, \mathbf{x}') = v(\mathbf{x} - \mathbf{x}')$. We will expand ψ in free particle functions. Thus

$$\psi(\mathbf{x}, t) = \sum_{\mathbf{k}} b_{\mathbf{k}}(t)\frac{e^{i\mathbf{k}\cdot\mathbf{x}}}{\sqrt{\Omega}} \quad (8\text{-}3)$$

Using this in Eq. 8-1 gives

$$H = \sum_{\mathbf{k}_1}\sum_{\mathbf{k}_1} b_{\mathbf{k}_1}^+ b_{\mathbf{k}_1}^+ \frac{\hbar^2 k_2^{\,2}}{2m}\int \frac{d^3x}{\Omega} e^{i(\mathbf{k}_2-\mathbf{k}_1)\cdot\mathbf{x}} + \sum_{\mathbf{k}_1}\sum_{\mathbf{k}_2}\sum_{\mathbf{k}_3}\sum_{\mathbf{k}_4} b_{\mathbf{k}_1}^+ b_{\mathbf{k}_2}^+ b_{\mathbf{k}_3} b_{\mathbf{k}_4} \times \int \frac{d^3x}{\Omega}\int\frac{d^3x'}{\Omega} e^{i(k_3-k_1)\cdot\mathbf{x}} e^{i(\mathbf{k}_4-\mathbf{k}_1)\cdot\mathbf{x}'} v(\mathbf{x} - \mathbf{x}') \quad (8\text{-}4)$$

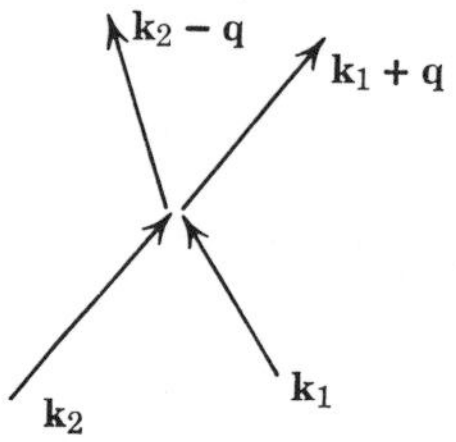

Figure 8-1

After making a change of variable in the last term and using Eq. 2-14, this can be put into the form

$$H = \sum_{\mathbf{k}} \frac{\hbar^2 k^2}{2m} b_{\mathbf{k}}^{+} b_{\mathbf{k}} + \sum_{\mathbf{k}_1} \sum_{\mathbf{k}_2} \sum_{\mathbf{q}} \bar{v}(\mathbf{q}) b_{\mathbf{k}_1+\mathbf{q}}^{+} b_{\mathbf{k}_2-\mathbf{q}}^{+} b_{\mathbf{k}_2} b_{\mathbf{k}_1} \tag{8-5a}$$

where

$$\bar{v}(\mathbf{q}) = \int \frac{d^3x}{\Omega}\, v(\mathbf{x}) e^{-i\mathbf{q}\cdot\mathbf{x}} \tag{8-5b}$$

is the Fourier transform of the interaction potential.

The interaction term in H can be represented by the diagram of Fig. 8-1. At the vertex particles or momenta $\hbar\mathbf{k}_1$ and $\hbar\mathbf{k}_2$ are destroyed by the destruction operators $b_{\mathbf{k}_1}$ and $b_{\mathbf{k}_2}$, and particles of momenta $h(\mathbf{k}_2 - \mathbf{q})$ and $\hbar(\mathbf{k}_1 - \mathbf{q})$ are created by the creation operators $b_{\mathbf{k}_1+\mathbf{q}}^{+}$ and $b_{\mathbf{k}_2-\mathbf{q}}^{+}$. The net result is a scattering of particles with an exchange of momentum $\hbar\mathbf{q}$. Momentum is conserved in the process. The amplitude for the scattering is $\bar{v}(\mathbf{q})$.

THE BOLTZMANN EQUATION FOR QUANTUM GASES BOSE-EINSTEIN AND FERMI-DIRAC DISTRIBUTIONS

Let us consider particles that interact through the potential $v(\mathbf{x})$. Let $N(\mathbf{k})$ be the number of particles of momentum $\hbar\mathbf{k}$. This number will change because of collisions among the particles. An equation for the rate of change of $N(\mathbf{k})$ may be written schematically as

$$\frac{\partial}{\partial t} N(\mathbf{k}) = \sum_{\mathbf{k}'} \sum_{\mathbf{q}} \left\{ \begin{array}{cc} \mathbf{k} & \mathbf{k}' \\ \nwarrow & \nearrow \\ \nearrow & \nwarrow \\ \mathbf{k}+\mathbf{q} & \mathbf{k}'-\mathbf{q} \end{array} \quad - \quad \begin{array}{cc} \mathbf{k}'-\mathbf{q} & \mathbf{k}+\mathbf{q} \\ \nwarrow & \nearrow \\ \nearrow & \nwarrow \\ \mathbf{k}' & \mathbf{k} \end{array} \right\} \tag{8-6}$$

We have added all of the processes that leave a particle with momentum $\hbar\mathbf{k}$ and subtracted all of the processes that remove a particle from the state with momentum $\hbar\mathbf{k}$. To get a mathematical equation from this we replace each diagram by the corresponding transition probability per unit time calculated by applying first order perturbation theory to the interaction term in

Eq. 8-4. Using Eqs. 4-23 and 4-24 we find

$$\frac{\partial}{\partial t} N(\mathbf{k}) = \sum_{\mathbf{k}'}\sum_{\mathbf{q}} \frac{2\pi}{\hbar} |\bar{v}(\mathbf{q})|^2 \, \delta\left[\frac{\hbar^2}{2m}(|\mathbf{k}+\mathbf{q}|^2 + |\mathbf{k}'-\mathbf{q}|^2 - k^2 - k'^2)\right]$$
$$\times \{N(\mathbf{k}+\mathbf{q})N(\mathbf{k}'-\mathbf{q})[1 \pm N(\mathbf{k})][1 \pm N(\mathbf{k}')]$$
$$- N(\mathbf{k})N(\mathbf{k}')[1 \pm N(\mathbf{k}+\mathbf{q})][1 \pm N(\mathbf{k}'-\mathbf{q})]\} \quad (8\text{-}7)$$

where the plus sign is to be used if the particles are Bosons, and the minus sign is to be used if the particles are Fermions. Note that when a particle is created at a vertex one gets a factor of $1 \pm N(\mathbf{k})$ in the square of the matrix element. If the state with momentum $\mathbf{k}$ is already occupied, then this factor is zero, so transitions of Fermions into occupied states is forbidden. On the other hand transitions of Bosons into occupied states is enhanced. Equation 8-7 is the quantum-mechanical generalization of the classical Boltzmann equation. Note that the scattering probability is proportional to $|\bar{v}(\mathbf{q})|^2$ which is the Born approximation result.

It can be seen almost by inspection that

$$\frac{\partial N(\mathbf{k})}{\partial t} = 0 \quad (8\text{-}8)$$

when

$$N(\mathbf{k}) = \frac{1}{Ce^{E(k)/T} \pm 1} \quad (8\text{-}9)$$

where $E(k) = \hbar^2k^2/2m$, and T is an energy and may be identified with the temperature of the system in energy units. The C is a normalization constant determined by

$$N_T = \sum_{\mathbf{k}} N(\mathbf{k}) = \Omega \int d^3k N(\mathbf{k})$$
$$= \Omega \int d^3k \frac{1}{Ce^{\hbar^2k^2/2mT} \pm 1} \quad (8\text{-}10)$$

where N_T is the total number of particles in the system. The plus sign is to be used in Eq. 8-9 if the particles are Fermions. The negative sign is to be used if they are Bosons. These distribution functions are called the Fermi-Dirac and Bose-Einstein distribution functions.

Problem 8-1. Show that Eq. 8-9 is an equilibrium solution of Eq. 8-7.

It is possible to prove an H-theorem using Eq. 8-7. For this purpose we need an expression for the entropy of a quantum gas. Landau and Lifschitz[51] give the entropy as

$$S = \pm K \sum_{\mathbf{k}} \{[1 \pm N(\mathbf{k})] \log [1 \pm N(\mathbf{k})] \pm N(\mathbf{k}) \log N(\mathbf{k})\} \quad (8\text{-}11)$$

where K is Boltzmann's constant. The upper sign applies to Bosons and the lower sign applies to Fermions. One can show that

$$\frac{dS}{dt} \geq 0 \tag{8-12}$$

is a consequence of Eq. 8-7. We outline the proof for Bosons; the proof for Fermions is similar. Differentiating Eq. 8-11 and then using 8-7 give

$$\begin{aligned}\frac{dS}{dt} &= K \sum_{\mathbf{k}} \frac{\partial N(\mathbf{k})}{\partial t} \{\log [N(\mathbf{k}) + 1] - \log N(\mathbf{k})\} \\ &= K \sum_{\mathbf{k}} \sum_{\mathbf{k}} \sum_{\mathbf{q}} \frac{2\pi}{\hbar} |\bar{v}(\mathbf{q})|^2 \, \delta\left[\frac{\hbar^2}{2m} (|\mathbf{k} + \mathbf{q}|^2 + |\mathbf{k}' - \mathbf{q}|^2 - k^2 - k'^2)\right] \\ &\quad \times \{N(\mathbf{k} + \mathbf{q})N(\mathbf{k}' - \mathbf{q})[1 + N(\mathbf{k})][1 + N(\mathbf{k}')] - N(\mathbf{k})N(\mathbf{k}') \\ &\quad \times [1 + N(\mathbf{k} + \mathbf{q})][1 + N(\mathbf{k}' - \mathbf{q})]\}\{\log [N(\mathbf{k}) + 1] - \log N(\mathbf{k})\}\end{aligned} \tag{8-13}$$

Next, one rewrites this equation making the change of variable $\mathbf{q} \to -\mathbf{q}$ and then letting $\mathbf{k} \to \mathbf{k} + \mathbf{q}$ and $\mathbf{k}' \to \mathbf{k}' - \mathbf{q}$; we call this our second equation. We do not write it down. Then, one gets a third equation by making the change of variable $\mathbf{k} \leftrightarrow \mathbf{k}'$, $\mathbf{q} \to -\mathbf{q}$ in Eq. 8-13. Further, one gets a fourth equation by making the change of variable $\mathbf{k} \leftrightarrow \mathbf{k}'$, $\mathbf{q} \to -\mathbf{q}$ in the second equation. Finally, using $|\bar{v}(-\mathbf{q})| = |\bar{v}(\mathbf{q})|$ and adding all four equations one obtains

$$\begin{aligned}4\frac{dS}{dt} &= \sum_{\mathbf{k}} \sum_{\mathbf{k}} \sum_{\mathbf{k}} \frac{2\pi}{\hbar} |\bar{v}(q)|^2 \, \delta\left[\frac{\hbar^2}{2m} (|\mathbf{k} + \mathbf{q}|^2 + |\mathbf{k}' - \mathbf{q}|^2 - \mathbf{k}^2 - \mathbf{k}'^2)\right] \\ &\quad \times \{N(\mathbf{k} + \mathbf{q})N(\mathbf{k}' - \mathbf{q})[1 + N(\mathbf{k})][1 + N(\mathbf{k}')] \\ &\quad - N(\mathbf{k})N(\mathbf{k}')[1 + N(\mathbf{k} + \mathbf{q})][1 + N(\mathbf{k}' - \mathbf{q})]\} \\ &\quad \times \{\log N(\mathbf{k} + \mathbf{q})N(\mathbf{k}' - \mathbf{q})[1 + N(\mathbf{k})][1 + N(\mathbf{k}')] \\ &\quad - \log N(\mathbf{k})N(\mathbf{k}')[1 + N(\mathbf{k} + \mathbf{q})][1 + N(\mathbf{k}' - \mathbf{q})]\}\end{aligned} \tag{8-14}$$

The product of the last two factors is of the form

$$\{x - y\}\{\log x - \log y\}$$

which is positive when $x > y$ and also when $x < y$; it vanishes when $x = y$. We conclude that Eq. 8-12 is true and that the equality sign holds only when Eq. 8-8 is true. This shows that the entropy increases monotonically and reaches its maximum value when the system attains its equilibrium distribution as given by Eq. 8-9.

The classical Boltzmann equation can be obtained from Eq. 8-7 by taking the classical limit $\hbar \to 0$. At the same time we let $\Omega \to \infty$. Also we assume that the gas is far from degeneracy so that $1 \pm N(\mathbf{k}) \simeq 1$. We define a velocity distribution $f(\mathbf{v})$ by

$$\sum_{\mathbf{k}} N(\mathbf{k}') \to \Omega \int d^3v' f(\mathbf{v}') \tag{8-15a}$$

and use

$$\hbar\mathbf{k} \to m\mathbf{v} \tag{8-15b}$$

$$\hbar\mathbf{q} \to m\mathbf{u} \tag{8-15c}$$

$$\sum_{\mathbf{q}} \to \Omega \int \frac{d^3q}{(2\pi)^3} = \frac{\Omega m^3}{\hbar^3(2\pi)^3} \int d^3u \tag{8-15d}$$

In this limit Eq. 8-7 becomes

$$\begin{aligned}\frac{\partial f_1(\bar{v})}{\partial t} = \int d^3v' \int d^3u \, \frac{m^3\Omega^2}{(2\pi)^2\hbar^4} \left| \bar{v}\left(\frac{m\mathbf{u}}{\hbar}\right)\right|^2 \\ \times \delta\left[\frac{m}{2}|\mathbf{v}+\mathbf{u}|^2 + \frac{m}{2}|\mathbf{v}'-\mathbf{u}|^2 - \frac{m}{2}v^2 - \frac{m}{2}v'^2\right] \\ \times \{f(\mathbf{v}+\mathbf{u})f(\mathbf{v}'-\mathbf{u}) - f(\mathbf{v})f(\mathbf{v}')\}\end{aligned} \tag{8-16}$$

Problem 8-2. Show that an equilibrium solution of Eq. 8-16 is

$$f(v) = Ce^{-mv^2/2T} \tag{8-17}$$

Problem 8-3. Show that when a quantum gas is far from degeneracy, Eq. 8-11 reduces to

$$S = -K \sum_{\mathbf{k}} N(\mathbf{k}) \log N(\mathbf{k}) \tag{8-18}$$

which is the classical definition of entropy.

Problem 8-4. Using the classical definition of entropy prove an H-theorem for Eq. 8-16.

THE DEGENERATE NEARLY PERFECT BOSE-EINSTEIN GAS

Let us examine Eq. 8-10 for the Bose-Einstein gas. Let

$$E(k) = \frac{\hbar^2k^2}{2m} = Tx \tag{8-19}$$

and

$$C = e^{x_0} \tag{8-20}$$

Then

$$\frac{N_T}{2\pi\Omega T^{3/2}}\left(\frac{\hbar}{2\pi}\right)^{3/2} = f(x_0) = \int_0^\infty \frac{x^{1/2}\,dx}{e^{x_0+x}-1} \tag{8-21}$$

We can evaluate $f(x_0)$ by a series expansion of the integrand and obtain

$$f(x_0) = \Gamma(\tfrac{3}{2}) \sum_{n=1}^{\infty} \frac{e^{-nx_0}}{n^{3/2}} \tag{8-22}$$

This is a monotonically decreasing function of x_0. For a sufficiently low value of T there is no value of x_0 for which Eq. 8-21 is satisfied. If one substitutes for N_T/Ω the density of liquid helium, one finds that the critical temperature is $T_c \simeq 3.2°\text{K}$. Below this value Eq. 8-21 cannot be satisfied.

The trouble lies in our replacement of the sum by the integral in Eq. 8-10. If we treat the zero energy state separately and write

$$N_T = \frac{1}{e^{x_0} - 1} + \Omega \int d^3k \frac{1}{e^{(E_0+E)/T} - 1} \tag{8-23}$$

where $E_0 = x_0 T$ then it can be shown that this has a solution for x_0 for every choice of N_T/Ω and T_0. As the temperature goes to zero more and more of the particles go into the zero energy state; finally, at $T = 0$ all of the particles are in the same state.

We saw in Chapter 2 that when a large number of particles were in the same state the field of which the particles are the quanta behaves classically. In Chapter 2 the particles were photons, but this conclusion should be true for any particles obeying Bose-Einstein statistics. Let us now consider a system of Bosons at $T = 0$ which is "slightly imperfect"; that is, we retain the interaction term in Eq. 8-5. In order to get a solvable model we treat $\bar{v}(\mathbf{q})$ as a constant that can be removed from the summation.

Now, the commutator

$$b_0 b_0^+ - b_0^+ b_0 = 1$$

of the operators of the $\mathbf{k} = 0$ state is very small in comparison with N, the eigenvalue of $b_0^+ b_0$, so in a sense these operators almost commute. This suggest that we treat b_0 and b_0^+ as C-numbers approximately equal to $\sqrt{N}$. In the interaction term of Eq. 8-5 there will be a zero order term

$$b_0^+ b_0^+ b_0 b_0 \simeq b_0^{\,4} \simeq N^2 \tag{8-24}$$

There are no first order terms containing one factor of b_q or $b_\mathbf{q}^+$ since these would not conserve momentum. The second order terms are

$$b_0^{\,2} \sum_{\mathbf{q} \neq 0} (b_\mathbf{q}^+ b_{-\mathbf{q}}^+ + b_\mathbf{q} b_{-\mathbf{q}} + 4 b_\mathbf{q}^+ b_\mathbf{q}) \tag{8-25}$$

To second order accuracy in Eq. 8-25 we can use $b_0^{\,2} = N$ but we need to do better in Eq. 8-24. We must use

$$b_0^{\,2} + \sum_{\mathbf{q} \neq 0} b_\mathbf{q}^+ b_\mathbf{q} = N \tag{8-26}$$

so that

$$b_0^4 = N^2 - 2N \sum_{\mathbf{q} \neq 0} b_{\mathbf{q}}^+ b_{\mathbf{q}} \tag{8-27}$$

As a result the Hamiltonian of Eq. 8-5 correct to second order in the "small" operators $b_{\mathbf{q}}$ and $b_{\mathbf{q}}^+$ is

$$H = \sum_{\mathbf{k}} \frac{\hbar k^2}{2m} b_{\mathbf{k}}^+ b_{\mathbf{k}} + N^2\bar{v} + N^2\bar{v} \sum_{\mathbf{k} \neq 0} (b_{\mathbf{k}}^+ b_{-\mathbf{k}}^+ + b_{\mathbf{k}} b_{-\mathbf{k}} + 2b_{\mathbf{k}}^+ b_{\mathbf{k}}) \tag{8-28}$$

We have reduced the Hamiltonian to a sufficiently simple form that now we can make a canonical transformation to new operators $a_{\mathbf{k}}$ and $a_{\mathbf{k}}^+$ which puts the Hamiltonian into the form

$$H = N^2\bar{v} + \sum_{\mathbf{k}} \varepsilon(\mathbf{k}) a_{\mathbf{k}}^+ a_{\mathbf{k}} \tag{8-29}$$

The appropriate transformation has the form

$$a_{\mathbf{k}} = \frac{b_{\mathbf{k}} + L_{\mathbf{k}} b_{-\mathbf{k}}^+}{\sqrt{1 - L_{\mathbf{k}}^2}} \tag{8-30a}$$

$$a_{\mathbf{k}}^+ = \frac{b_{\mathbf{k}}^+ + L_{\mathbf{k}} b_{-\mathbf{k}}}{\sqrt{1 - L_{\mathbf{k}}^2}} \tag{8-30b}$$

where $L_{\mathbf{k}}$ is a real number whose value is still to be determined. It is readily checked that

$$[a_{\mathbf{k}}, a_{\mathbf{k}}^+]_- = \delta_{\mathbf{k},\mathbf{k}'} \tag{8-31a}$$

$$[a_{\mathbf{k}}, a_{\mathbf{k}'}]_- = [a_{\mathbf{k}}^+, a_{\mathbf{k}}^+]_- = 0 \tag{8-31b}$$

follows from the commutation relations for $b_{\mathbf{k}}$ and $b_{\mathbf{k}}^+$, whatever the value of $L_{\mathbf{k}}$. The inverse transformation of Eq. 8-30 is

$$b_{\mathbf{k}} = \frac{a_{\mathbf{k}} - L_{\mathbf{k}} a_{-\mathbf{k}}^+}{\sqrt{1 - L_{\mathbf{k}}^2}} \tag{8-32a}$$

$$b_{\mathbf{k}}^+ = \frac{a_{\mathbf{k}}^+ - L_{\mathbf{k}} a_{-\mathbf{k}}}{\sqrt{1 - L_{\mathbf{k}}^2}} \tag{8-32b}$$

When these are used in Eq. 8-28, it is found that H reduces to the form of Eq. 8-29 if $L_{\mathbf{k}}$ is chosen to be

$$L_{\mathbf{k}} = \frac{1}{2N\bar{v}}\left\{\varepsilon(\mathbf{k}) - \frac{\hbar^2 k^2}{2m} - 2\bar{v}N\right\} \tag{8-33}$$

where

$$\varepsilon(\mathbf{k}) = \sqrt{\frac{2\bar{v}N\hbar^2k^2}{m} + \left(\frac{\hbar^2k^2}{2m}\right)^2} \tag{8-34}$$

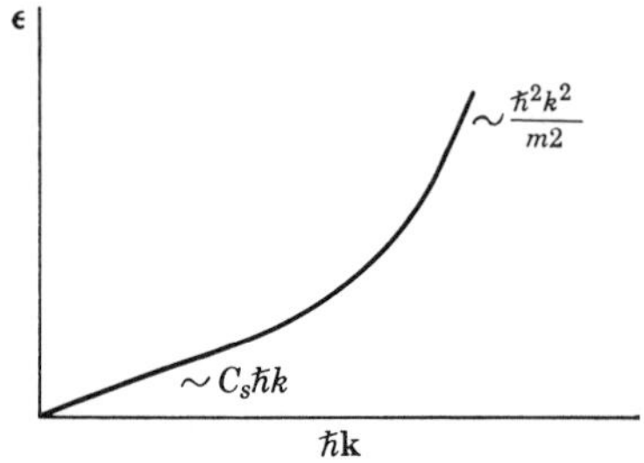

Figure 8-2

Equation 8-29 describes a system of quanta of momentum $\hbar\mathbf{k}$ whose energies are given by $\varepsilon(\mathbf{k})$. The operators $a_{\mathbf{k}}^{+}$ and $a_{\mathbf{k}}$ create and destroy these quanta. Note that for small $\mathbf{k}$,

$$\varepsilon(\mathbf{k}) \simeq k\sqrt{\frac{2\bar{v}N\hbar^2}{m}} = C_s k\hbar \tag{8-35}$$

where $C_s = \sqrt{2\bar{v}N/m}$ is a velocity. It may be interpreted as the velocity of sound in the degenerate gas. These long wavelength excitations are called phonons. In the short wavelength (high momentum) limit, Eq. 8-34 becomes

$$\varepsilon(k) \simeq \frac{\hbar^2 k^2}{2m} \tag{8-36}$$

This is the energy-momentum relation with which we started. In this limit the excitations behave like noninteracting particles. The energy-momentum relation is sketched in Fig. 8-2.

The phonon is a good example of a "quasi particle." In a certain approximation the interacting particles of the gas behave like a gas of different particles, the quasi particles, which do not interact.

SUPERFLUIDITY

Consider an impurity atom moving through a zero temperature fluid with an energy-momentum relation such as that shown in Fig. 8-2. The only way the impurity atom can lose energy is for it to create an excitation in the fluid. (At nonzero temperature there will already be excitations present in the fluid which can scatter on the impurity atom and exchange energy with it, but at zero temperature there will be no excitations present.) If we suppose that the impurity atom initially has momentum $\hbar\mathbf{q}$ and emits an excitation of momentum $\hbar\mathbf{k}$ then conservation of momentum and energy gives

$$\frac{\hbar^2 q^2}{2m} = \frac{\hbar^2}{2m}|\mathbf{q} - \mathbf{k}|^2 + \varepsilon(\mathbf{k}) \tag{8-37}$$

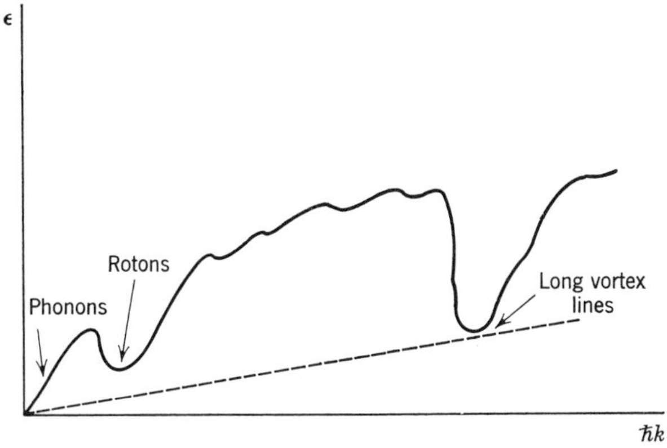

Figure 8-3

from which

$$\cos\theta = \frac{\varepsilon/\hbar k}{v} + \frac{\hbar k/2m}{v} \tag{8-38}$$

where θ is the angle between q and k and $v = \hbar q/m$ is the initial velocity of the impurity atom. For phonons $\varepsilon/\hbar k > C_s$, so that v must be greater than C_s for a phonon to be emitted. Impurity atoms moving with a velocity less than a critical velocity (which in this case is C_s) can not lose energy to the fluid.

We can also look at this in a frame of reference in which the impurity atom is stationary and the fluid flows past it. There will be no frictional force unless the critical velocity is exceeded. This result should also be true if the impurity atom is replaced by a rough place on the wall of the tube through which the fluid flows.

In liquid helium the critical velocity is much less than the velocity of sound. It is suspected that the energy-momentum relation must be like that of Fig. 8-3. The critical velocity is determined by the slope of the dotted line shown. The large momentum excitations responsible for the critical velocity may be long vortex lines.

Problem 8-5. Do the following experiment. Fill the kitchen sink with water. Now, move some thin object such as a knife blade through the water, slowly at first, and then increase the speed a little more each time you do it. Note that at low speeds there is laminar flow about the object. Above a critical speed the character of the flow changes. Why?

9

Quasi Particles in Plasmas and Metals

In the preceding chapter it was seen how a system of interacting particles could behave, in a certain approximation, as a system of noninteracting quasi particles. We discuss two other quasi particles in this chapter. We suppose that the system under consideration consists of a collection of electrons and ions that has an overall electrical neutrality. Such a system is called a plasma. It is assumed to be isotropic and homogeneous. In some respects this is not a bad approximation to a metal, but of course, properties related to the periodicity of a true solid are missing from this model.

From the beginning we make the self-consistent field approximation. That is, we assume that the particles interact with an electrostatic potential $\phi(\mathbf{x}, t)$ which in turn is to be calculated from the "average" charge density in the plasma. Just how this average is to be calculated will be made clear presently. The Hamiltonian for the system may be written as

$$H = \sum_s \int d^3x \psi_s^+ \left[\frac{\hbar^2}{2m_s} \nabla^2 + e_s \phi \right] \psi_s = H_0 + H_I \tag{9-1}$$

where s ranges over the species of particles in the plasma (usually, electrons and ions) and H_I contains the terms involving ϕ. We expand ψ_s and ψ_s^+ in free particle wave functions; thus

$$\psi_s = \sum_{\mathbf{q}} b_{s\mathbf{q}} \frac{e^{i\mathbf{q}\cdot x}}{\sqrt{\Omega}} \tag{9-2}$$

In the usual way H_0 becomes

$$H_0 = \sum_s \sum_{\mathbf{q}} \frac{\hbar^2 q^2}{2m_s} b_{s\mathbf{q}}^+ b_{s\mathbf{q}} \tag{9-3}$$

The interaction Hamiltonian becomes

$$H_I = e_s \sum_s \sum_{\mathbf{q}_1} \sum_{\mathbf{q}_2} b_{s\mathbf{q}_1}^+ b_{s\mathbf{q}_2} \bar{\phi}(\mathbf{q}_2 - \mathbf{q}_1) \tag{9-4a}$$

Where

$$\bar{\phi}(\mathbf{q}_2 - \mathbf{q}_1) = \int \frac{d^3x}{\Omega}\, e^{i(\mathbf{q}_2-\mathbf{q}_1)\cdot\mathbf{x}}\phi(\mathbf{x}) \tag{9-4b}$$

is the Fourier transform of $\phi(\mathbf{x})$.

The Heisenberg equations of motion may be used to calculate the rate of change of any operator constructed from $b_{s\mathbf{q}}$ and $b_{s\mathbf{q}}^+$. We are particularly concerned with the operator $b_{s\mathbf{q}'}^+ b_{s\mathbf{q}}$. For it we find

$$-\frac{\hbar}{i}\frac{\partial}{\partial t}\, b_{s\mathbf{q}'}^+ b_{s\mathbf{q}} = [(H_0 + H_I),\, b_{s\mathbf{q}'}^+ b_{s\mathbf{q}}]_- \tag{9-5}$$

The equations of motion are found to be the same whether we use the commutation relations, Eq. 4-23, for Bosons or Eq. 4-24 for Fermions. In either case we get

$$\frac{\partial}{\partial t}\, b_{s\mathbf{q}'}^+ b_{s\mathbf{q}} = +\frac{i}{\hbar}(E_{s\mathbf{q}'} - E_{s\mathbf{q}})b_{s\mathbf{q}'}^+ b_{s\mathbf{q}}$$

$$+ \frac{ie_s}{\hbar}\sum_{\mathbf{p}} \{\bar{\phi}(\mathbf{p} - \mathbf{q})b_{s\mathbf{q}'}^+ b_{s\mathbf{p}} - \bar{\phi}(\mathbf{q}' - \mathbf{p})b_{s\mathbf{p}}^+ b_{s\mathbf{q}}\} \tag{9-6}$$

We now define a function $F_s(\mathbf{q}', \mathbf{q}, t)$ which we call the distribution function for particles of species s; it is defined as

$$F_s(\mathbf{q}', \mathbf{q}, t) = \sum_{\alpha} P_\alpha \langle\alpha|\, b_{s\mathbf{q}'}^+ b_{s\mathbf{q}'}\, |\alpha\rangle \tag{9-7}$$

Where the states of the system are the $|\alpha\rangle$'s and P_α is the probability of finding the system in the state $|\alpha\rangle$. The equation of motion for F_s is found from Eq. 9-6 to be

$$\frac{\partial}{\partial t} F_s(\mathbf{q}', \mathbf{q}, t) = \frac{i}{\hbar}(E_{s\mathbf{q}'} - E_{s\mathbf{q}})F_s(\mathbf{q}', \mathbf{q}, t)$$

$$+ \frac{ie_s}{\hbar}\sum_{\mathbf{p}} \{\bar{\phi}(\mathbf{p} - \mathbf{q})F_s(\mathbf{q}', \mathbf{p}, t) - \bar{\phi}(\mathbf{q}' - \mathbf{p})F_s(\mathbf{p}, \mathbf{q}, t)\} \tag{9-8}$$

We digress briefly to discuss the meaning and the usefulness of the quantum mechanical operator we have just defined. In Chapter 4 we defined a number density operator by $n = \psi^+\psi$. If we average this by the averaging process of Eq. 9-7 we obtain

$$\langle n_s(\mathbf{x}, t)\rangle = \sum_{\alpha} P_\alpha\langle\alpha|\, \psi_s^+(\mathbf{x}, t)\psi_s(\mathbf{x}, t)\, |\alpha\rangle \tag{9-9}$$

Using Eq. 9-2, this may be written as

$$\langle n_s(\mathbf{x}, t)\rangle = \sum_{\mathbf{q}}\sum_{\mathbf{p}} \langle b_{s\mathbf{p}}^+ b_{s\mathbf{p}+\mathbf{q}}\rangle \frac{e^{i\mathbf{q}\cdot\mathbf{x}}}{\Omega}$$

$$= \sum_{\mathbf{q}}\sum_{\mathbf{p}} F_s(\mathbf{p}, \mathbf{p} + \mathbf{q}, t)\frac{e^{i\mathbf{q}\cdot\mathbf{x}}}{\Omega} \tag{9-10}$$

This suggests that we define a coordinate and momentum space distribution function by

$$F_s(\mathbf{x}, \mathbf{p}, t) = \sum_{\mathbf{q}} F_s(\mathbf{p}, \mathbf{p} + \mathbf{q}, t)\frac{e^{i\mathbf{q}\cdot\mathbf{x}}}{\Omega} \tag{9-11}$$

for then

$$\langle n_s(\mathbf{x}, t)\rangle = \sum_{\mathbf{p}} F_s(\mathbf{x}, \mathbf{p}, t) \tag{9-12}$$

Furthermore,

$$\langle n_s(\mathbf{p}, t)\rangle = \int d^3x F_s(\mathbf{x}, \mathbf{p}, t) = \langle b_{s\mathbf{p}}^+ b_{s\mathbf{p}}\rangle \tag{9-13}$$

is the momentum distribution function for species s. Equations 9-12 and 9-13 are the properties we would expect of a distribution function. If F_s were a classical function then we could interpret $F_s(\mathbf{x}, \mathbf{p}, t)d^3x\, d^3p$ as the probable number of particles with coordinates in d^3x and momentum in d^3p. Such a description is not possible in quantum mechanics; still the quantum-mechanical distribution function is in many ways analogous to a classical distribution function.

Equation 9-10 can be used to calculate the charge density in the plasma. This is then used in Poisson's equation to obtain

$$\begin{aligned}\nabla^2\phi &= -\sum_s 4\pi c_s\langle n(\mathbf{x}, t)\rangle \\ &= -\sum_s\sum_{\mathbf{q}}\sum_{\mathbf{p}} \frac{4\pi e_s}{\Omega} F_s(\mathbf{p}, \mathbf{p} + \mathbf{q}, t)e^{i\mathbf{p}\cdot\mathbf{x}}\end{aligned} \tag{9-14}$$

In this way the potential is made "self-consistent." It is clear that an approximation has been made in replacing the true charge density with the average charge density. This is known as the Hartree approximation in the theory of atomic structure. The coupled equations Eqs. 9-8 and 9-14 are the quantum-mechanical analogs of the Vlasov equations which are well known to plasma physicists.

Next, we look for small oscillations about an equilibrium in which the charge density and the potential, ϕ, vanish. We write

$$F_s(\mathbf{q}', \mathbf{q}, t) = F_{s0}(\mathbf{q})\,\delta_{\mathbf{q},\mathbf{q}'} + F_{s1}(\mathbf{q}', \mathbf{q})e^{-i\omega t} \tag{9-15}$$

and treat F_{s1} and ϕ as small quantities whose products may be neglected. Equation 9-8 becomes

$$F_{s1}(\mathbf{q}', \mathbf{q}) = \frac{e_s}{\hbar}\,\bar{\phi}(\mathbf{q}' - \mathbf{q})\,\frac{[F_{s0}(\mathbf{q}') - F_{s0}(\mathbf{q})]}{\omega - \nu_s(\mathbf{q}', \mathbf{q})} \tag{9-16a}$$

where

$$\nu_s(\mathbf{q}', \mathbf{q}) = (E_{s\mathbf{q}'} - E_{s\mathbf{q}})/\hbar \tag{9-16b}$$

This may be used in Eq. 9-14 to obtain

$$\nabla^2\phi = \sum_s \sum_{\mathbf{q}} \sum_{\mathbf{p}} \frac{4\pi e_s^{\,2}}{\hbar\Omega}\, \bar{\phi}(-\mathbf{q})\, \frac{\{F_{s0}(\mathbf{p}) - F_{s0}(\mathbf{p}+\mathbf{q})\}}{\omega - \nu_s(\mathbf{p}, \mathbf{p}+\mathbf{q})}\, e^{i\mathbf{q}\cdot\mathbf{x}} \tag{9-17}$$

Writing

$$\phi(\mathbf{x}) = \sum_{\mathbf{q}} \bar{\phi}(-\mathbf{q}) e^{i\mathbf{q}\cdot\mathbf{x}} \tag{9-18}$$

we see that $\bar{\phi}(-\mathbf{q})$ must satisfy

$$\varepsilon(\mathbf{q}, \omega)\bar{\phi}(-\mathbf{q}) = 0 \tag{9-19}$$

where

$$\varepsilon(\mathbf{q}, \omega) = 1 + \sum_s \sum_{\mathbf{p}} \frac{4\pi e_s^{\,2}}{q^2\hbar\Omega}\, \frac{F_{s0}(\mathbf{p}) - F_{s0}(\mathbf{p}-\mathbf{q})}{\omega - \nu_s(\mathbf{p}, \mathbf{p}-\mathbf{q})} \tag{9-20}$$

is called the dielectric function of the plasma.

From Eq. 9-19 it is seen that either $\bar{\phi}(\mathbf{q}) = 0$ or

$$\varepsilon(\mathbf{q}, \omega) = 0 \tag{9-21}$$

This equation may be solved for ω to obtain the one or more frequencies with which a wave of wave number $\mathbf{q}$ can propagate. Before discussing the solution of this equation it is convenient to replace $F_{s0}(\mathbf{p})$ by the corresponding velocity distribution function $f_{s0}(\mathbf{v})$ where $\mathbf{v} = \hbar\mathbf{p}/m$. Also we let the volume of the system become infinite and use

$$\sum_{\mathbf{p}} \to \Omega \int d^3v \tag{9-22}$$

to obtain

$$\varepsilon(\mathbf{q}, w) = 1 + \sum_s \frac{4\pi e_s^{\,2}}{\hbar q^2} \int d^3v\, \frac{f_{s0}(\mathbf{v}) - f_{s0}(\mathbf{v} - \hbar\mathbf{q}/m_s)}{\omega - \mathbf{q}\cdot\mathbf{v} + \hbar q^2/2m_s} \tag{9-23}$$

The classical dielectric function of a plasma may be obtained by taking the $\hbar \to 0$ limit; it is

$$\varepsilon_c(\mathbf{q}, \omega) = 1 + \sum_s \frac{4\pi e_s^{\,2}}{m_s q^2} \int d^3v\, \frac{\mathbf{q}\cdot\partial f_{s0}/\partial\mathbf{v}}{\bar{\omega} - \mathbf{q}\cdot\mathbf{v}} \tag{9-24}$$

There is a little difficulty about ε as we have derived it which must be removed before we can proceed. There is a value of $\mathbf{v}$ for which the denominator of the integrand in $\varepsilon(\mathbf{q}, \omega)$ vanishes; the integrals are improper. This difficulty and its interpretation has given rise to a considerable body of literature. Landau[52] first called attention to this problem and showed how it could be resolved by treating the problem as an initial value problem and using Laplace transforms. In Landau's treatment the frequency ω is replaced by the Laplace transform parameter which has a positive imaginary part. This removes the singularity from the real axis and makes the integrals proper. Values of $\varepsilon(\mathbf{q}, \omega)$ for other values of ω are then found by analytic continuation. We shall follow Landau's prescription to the extent of replacing ω by $\omega + i\eta$. Then we obtain $\varepsilon(\mathbf{q}, \omega)$ for real ω by taking the limit $\eta \to 0+$.

We may divide ε into a real and an imaginary part (for real ω) by using the Plemelj formula

$$\frac{1}{x + i\eta} \xrightarrow[\eta \to 0+]{} P\frac{1}{x} - i\pi\delta(x) \tag{9-25}$$

where P indicates that a principal part is to be taken in subsequent integrations. We obtain

$$\varepsilon(\mathbf{q}, \omega) = \varepsilon_1(\mathbf{q}, \omega) + i\varepsilon_2(\mathbf{q}, \omega) \tag{9-26a}$$

$$\varepsilon_1(\mathbf{q}, \omega) = 1 + \sum_s \frac{4\pi e_s^{\,2}}{m_s q^2} P \int d^3v \frac{f_{s0}(\mathbf{v}) - f_{s0}\left(\mathbf{v} - \dfrac{\hbar\mathbf{q}}{ms}\right)}{\omega - \mathbf{q}\cdot\mathbf{v} + \dfrac{\hbar^2 q^2}{2m}} \tag{9-26b}$$

$$\varepsilon_2(\mathbf{q}, \omega) = -\sum_s \frac{4\pi^2 e_s^{\,2}}{m_s q^2} \int d^3v [f_{s0}(\mathbf{v}) - f_{s0}(\mathbf{v} - \hbar\mathbf{q}/m_s)]\, \delta\left[\omega - \mathbf{q}\cdot\mathbf{v} + \frac{\hbar^2 q^2}{2m_s}\right] \tag{9-26c}$$

Generally, the roots of $\varepsilon(\mathbf{q}, \omega) = 0$ are complex indicating that the waves decay or grow exponentially. It may be shown that if $f_{s0}(\mathbf{v})$ is a monotonically decreasing function of $v = |\mathbf{v}|$, then the roots have a negative imaginary part indicating that the waves are damped. This is always the case in thermal equilibrium. If the plasma is far from thermal equilibrium, it is possible to have roots with a positive imaginary part. Such a wave would grow exponentially; the plasma is said to be unstable.

There is a very useful formula for finding the imaginary part of ω when this imaginary part is small and is due to ε_2. Let us write

$$\omega = \Omega + i\gamma \tag{9-27}$$

and assume that both γ and ε_2 are small quantities whose product is negligible. Then writing

$$\varepsilon(\mathbf{q}, \Omega + i\gamma) = 0 \simeq \varepsilon_1(\mathbf{q}, \Omega) + i\gamma \frac{\partial\varepsilon_1(\mathbf{q}, \Omega)}{\partial\Omega} + i\varepsilon_2(\mathbf{q}, \Omega) \tag{9-28}$$

and equating real and imaginary parts to zero gives

$$\varepsilon_1(\mathbf{q}, \Omega) = 0 \tag{9-29a}$$

and

$$\gamma = -\frac{\varepsilon_2(\mathbf{q}, \Omega)}{(\partial\varepsilon_1/\partial\Omega)(\mathbf{q}, \Omega)} \tag{9-29b}$$

The real part of ω is given by the first of these equations, and the imaginary part is given by the second. If the roots of Eq. 9-29a are complex instead of real then this method fails.

PLASMONS AND PHONONS

In order to simplify the calculations of this section we neglect the quantum corrections to the frequency and use the classical dielectric function, Eq. 9-24. We assume that the distribution functions for ions and electrons are degenerate Fermi-Dirac distributions; thus

$$f_{s0}(v) = \begin{cases} \dfrac{3n}{4\pi v_{fs}^{3}} & v < v_{fs} \\ 0 & v > v_{fs} \end{cases} \tag{9-30}$$

where v_{fs} is the Fermi velocity of particles of species s; it is given by

$$v_{fs} = \frac{\hbar}{m_s}\left(\frac{3n}{4\pi}\right)^{1/3} \tag{9-31}$$

where n is the particle density and is assumed to be the same for electrons and ions.

Note that

$$\mathbf{q}\cdot\frac{\partial f_{s0}}{\partial \mathbf{v}} = -\frac{\mathbf{q}\cdot\mathbf{v}}{v}\left(\frac{3n}{4\pi v_{fs}^{3}}\right)\delta(v - v_{fs}) \tag{9-32}$$

This makes the integrals rather easy to do. We find

$$\begin{aligned}\varepsilon(\mathbf{q}, \omega) &= 1 + \sum_s \frac{4\pi e_s^2}{m_s q^2}\int d^3v\,\frac{\mathbf{q}\cdot\partial f_{s0}/\partial\mathbf{v}}{\omega - \mathbf{q}\cdot\mathbf{v} + i\eta} \\ &= 1 + \sum_s \frac{3}{2}\frac{\omega_{ps}^2}{q^2 v_{fs}^2}\left\{2 - z_s \log\left|\frac{1 + z_s}{1 - z_s}\right| + i\pi z_s u(z_s)\right\}\end{aligned} \tag{9-33}$$

where we have let

$$z_s = \frac{\omega}{q v_{fs}} \tag{9-34a}$$

$$\omega_{ps} = (4\pi n e_s^2/m_s)^{1/2} \tag{9-34b}$$

and

$$u(z) = \begin{cases} 1 & |z| < 1 \\ 0 & |z| > 1 \end{cases} \tag{9-34c}$$

In searching for roots of $\varepsilon(\mathbf{q}, \omega) = 0$, let us first assume that $z_s \gg 1$ for both electrons and ions. (If it is true for electrons then it is necessarily true for ions since $v_{fi} = v_{fe}(m_e/m_i)$.) Using

$$2 - z \log \left| \frac{1+z}{1-z} \right| \simeq -\frac{2}{3}\frac{1}{z^2} - \frac{2}{5}\frac{1}{z^4} \tag{9-35}$$

we obtain

$$\varepsilon(\mathbf{q}, \omega) \simeq 1 - \frac{\omega_{pe}^2}{\omega^2}\left(1 + \frac{m_e}{m_i}\right) - \frac{3}{5}\frac{\omega_{pe}^2 v_{fe}^2(1 + m_e^3/m_i^3)q^2}{\omega^4} \tag{9-36}$$

Setting this equal to zero and solving approximately for ω gives

$$\omega^2 \simeq \omega_{pe}^2\left(1 + \frac{m_e}{m_i}\right) + \tfrac{3}{5}v_{fe}^2 q^2 \tag{9-37}$$

These oscillations are plasma oscillations. Their quanta are called plasmons. Their frequency is nearly equal to the electron plasma frequency ω_{pe}, but the motion of the ions modifies this by the factor $(1 + m_e/m_i)$. There is also a thermal correction given by the term $v_{fe}^2 q^2$. These waves are undamped since $\varepsilon_2(\mathbf{q}, \omega)$ vanishes for $z_s > 1$ according to Eq. 9-33.

There is also a solution of $\varepsilon(\mathbf{q}, \omega) = 0$ with $z_i \gg 1$ but $z_e \ll 1$. (This is possible since $z_e/z_i = m_e/m_i$). We make the approximations

$$2 - z_i \log \left| \frac{1+z_i}{1-z_i} \right| \simeq -\frac{2}{3z_i^2} \tag{9-38a}$$

$$2 - z_e \log \left| \frac{1+z_e}{1-z_e} \right| \simeq 2 \tag{9-38b}$$

Equation 9-33 becomes

$$\varepsilon(\mathbf{q}, \omega) = 1 - \frac{\omega_{pi}^2}{\omega^2} + 3\frac{\omega_{pe}^2}{q^2 v_{fe}^2} + i\frac{3\pi}{2}\frac{\omega_{pe}^2\omega}{q^3 v_{fe}^3} \tag{9-39}$$

Treating the last term as a small quantity and solving approximately for ω gives

$$\omega = \Omega\left(1 - \frac{i\Omega}{2qv_{fe}}\frac{(\omega_{pe}/qv_{fe})^2}{1 + 3\omega_{pe}^2/q^2v_{fe}^2}\right) \tag{9-40}$$

where

$$\Omega = \frac{\omega_{pi}}{\sqrt{1 + 3\omega_{pe}/q^2 v_{fe}^2}} \tag{9-41}$$

For very long wavelengths the frequency is given by

$$\omega \simeq \Omega = \frac{v_{fe}}{\sqrt{3}}\left(\frac{\omega_{pi}}{\omega_{pe}}\right)q = \frac{v_{fe}}{\sqrt{3}}\left(\frac{m_e}{m_i}\right)^{1/2} q \tag{9-42}$$

This is the frequency-wave number relation expected for a sound wave with velocity

$$\frac{\omega}{q} = \frac{v_{fe}}{\sqrt{3}}\left(\frac{m_e}{m_i}\right)^{1/2} \tag{9-43}$$

Actually, the velocity of sound given by Eq. 9-43 is in rather good agreement with the velocity of longitudinal sound waves observed in metals. The agreement is within 20% for the alkali metals. The quanta of these low frequency waves are called phonons. In the long wavelength limit Eq. 9-40 gives

$$\omega \simeq \Omega\left(1 - \frac{i}{3\sqrt{3}}\left(\frac{m_e}{m_i}\right)^{1/2}\right) \quad (9\text{-}44$$

indicating a weak damping of the phonons.

In concluding this section we will remark that the results are not changed much if the degenerate Fermi-Dirac distributions are replaced by Maxwellian distributions. The Fermi velocity v_{fs} is replaced by the thermal velocity $v_{ts} = (2T/m_s)^{1/2}$ and some numerical coefficients of order unity are changed slightly. The biggest change is in the damping of the waves. This is discussed further in the next section.

Problem 9-1. Use the dielectric function given in Eq. 9-33 to calculate the electrostatic potential about a stationary charge Q immersed in the plasma.

Problem 9-2. Assume the particles have Maxwellian distributions and use Eq. 9-24 to calculate the classical dielectric function.

LANDAU DAMPING IN PLASMAS AND CHARACTERISTIC ENERGY LOSSES IN SOLIDS

We now examine more carefully the damping of waves in a plasma. For that purpose it is useful to return to Eq. 9-20, replace ω by $\omega + i\eta$, and use Eq. 9-25 to obtain the imaginary part of $\varepsilon(\mathbf{q}, \omega)$ in the form

$$\varepsilon_2(\mathbf{q}, \omega) = \sum_s \sum_{\mathbf{p}} 4\pi^2 e_s^2 [F_{s0}(\mathbf{p}) - F_{s0}(\mathbf{p} - \mathbf{q})] \cdot \delta[h\omega - E_{s\mathbf{p}} + E_{s\mathbf{p}-\mathbf{q}}] \tag{9-45}$$

Multiplying numerator and denominator of Eq. 9-29b by $|\mathbf{E}|^2(\omega/4\pi)$ gives

$$\gamma = +\frac{P}{2W} \tag{9-46}$$

where

$$P = |\phi|^2 \pi \sum_s \sum_{\mathbf{p}} \frac{\omega e_s^2}{\Omega} [F_{s0}(\mathbf{p}) - F_{s0}(\mathbf{p} - \mathbf{q})] \cdot \delta(\hbar\omega - E_{sp} + E_{s\mathbf{p}-\mathbf{q}}) \tag{9-47a}$$

and

$$W = \frac{|\mathbf{E}|^2}{8\pi} \frac{\partial}{\partial \omega} (\omega \varepsilon_1(\mathbf{q}, \omega))_{\omega=\Omega} \tag{9-47b}$$

Now, W may be interpreted as the energy density of a longitudinal wave with electric field $\mathbf{E} = -i\mathbf{q}\phi$, since by Eq. 3-50 it is corrected by just the right factor to take into account the energy of the oscillating particles of the plasma. Since the energy of a wave is proportional to the square of the amplitude, we expect W to decay as $e^{+2\gamma t}$. Also, we can equate dW/dt to the rate of transfer of energy per unit volume from the particles of the plasma to the wave which we shall call P. Thus we get

$$\frac{1}{W}\frac{dW}{dt} = 2\gamma = \frac{P}{W} \tag{9-48}$$

as in Eq. 9-39. Now, P is just the rate of energy transfer per unit volume calculated by using the Fermi golden rule and Eq. 9-4. This rate is the difference between the rate of emission of quasi particles and absorption of quasi particles.

We can put the theory of plasmons and phonons on a more formal basis if we quantize the longitudinal field in a plasma in much the same way as we did the transverse field in Chapter 2. Let us write

$$\phi(\mathbf{x}, t) = \sum_{\mathbf{k}\sigma} \left| \frac{4\pi\hbar\Omega_{\mathbf{k}\sigma} e_s^{\,2}}{\Omega k^2 \left(\frac{\partial}{\partial \omega} \omega\varepsilon_1\right)_{\Omega_{\mathbf{k}\sigma}}} \right|^{1/2} \{a_{\mathbf{k}\sigma} e^{i(\mathbf{k}\cdot\mathbf{x} - \Omega_{k\sigma} t)} + a_{\mathbf{k}\sigma}^{+} e^{-i(\mathbf{k}\cdot\mathbf{x}) - \Omega_{\mathbf{k}\sigma} t)}\} \tag{9-49}$$

where $\Omega_{\mathbf{k}\sigma}$ are the roots of $\varepsilon_1(\mathbf{k}, \omega) = 0$. There may be more than one root for each wave vector $\mathbf{k}$ so that we must distinguish between them by the subscript σ. We now calculate the energy in the electric field

$$U = \frac{1}{8\pi} \int d^3\mathbf{x}\, \langle E^2(\mathbf{x}, t)\rangle \tag{9-50}$$

where the angle brackets indicate a time average over a period which is much longer than a period of oscillation. We find

$$U = \sum_{\mathbf{k},\sigma} \frac{\hbar\Omega_{\mathbf{k}\sigma}}{\left(\frac{\partial}{\partial\omega}\omega\varepsilon_1\right)_{\Omega_{\mathbf{k}\sigma}}} a_{\mathbf{k}\sigma}^{+} a_{\mathbf{k}\sigma} \tag{9-51}$$

Now, correcting each term by the factor

$$\left(\frac{\partial}{\partial\omega}\omega\varepsilon_1\right)_{\Omega_{\mathbf{k}\sigma}}$$

to take into account the energy of the particles gives for the total energy

$$H = \sum_{\mathbf{k}\sigma} \hbar\Omega_{\mathbf{k}\sigma} a^{+}_{\mathbf{k}\sigma} a_{\mathbf{k}\sigma} \tag{9-52}$$

This may be taken to be the Hamiltonian of the system of quasi particles. We interpret $a_{\mathbf{k}\sigma}{}^{+}$ and $a_{\mathbf{k}\sigma}$ as creation and annihilation operators of quasi particles of type σ, momentum $\hbar\mathbf{k}$, and energy $\hbar\Omega_{\mathbf{k}\sigma}$.

The interaction Hamiltonian may be obtained from the term containing ϕ in Eq. 9-1. Thus

$$H_I = \sum_s e_s \int d^3x \psi_s^{+} \phi \psi_s \tag{9-53}$$

Substituting the expansions Eqs. 9-2 and 9-42 and carrying out the integration give

$$H_I = \sum_{\mathbf{k},\sigma} \sum_{\mathbf{q},s} \left[\frac{4\pi e_s{}^2 \hbar\Omega_{\mathbf{k}\sigma}}{\Omega k^2 \left(\dfrac{\partial}{\partial\omega}\, \omega\varepsilon_1 \right)_{\Omega_{\mathbf{k}\sigma}}} \right]^{1/2} \{ b^{+}_{s\mathbf{q}+\mathbf{k}} b_{s\mathbf{q}} a_{\mathbf{k}\sigma} + b^{+}_{s\mathbf{q}} b_{s\mathbf{q}+\mathbf{k}} a^{+}_{\mathbf{k}\sigma} \} \tag{9-54}$$

The terms in Eq. 9-47 may be represented by the diagrams of Fig. 9-1.

We can use H_I in the Fermi golden rule to calculate the rate of change of $N_\sigma(\mathbf{k})$ the number of quasi particles of type σ of momentum $h\mathbf{k}$. Schematically, we write

$$\frac{\partial}{\partial t} N_\sigma(\mathbf{k}) = \sum_{\mathbf{q},s} \left\{ \begin{array}{ccc} \mathbf{k},\sigma & & s,\mathbf{q} \\ \nwarrow & \!\!\searrow\!\nearrow & \\ & \uparrow & \\ & s,\mathbf{q}+\mathbf{k} & \end{array} \quad - \quad \begin{array}{ccc} & s,\mathbf{q}+\mathbf{k} & \\ & \uparrow & \\ & \nearrow\nwarrow & \\ s,\mathbf{q} & & \mathbf{k}\sigma \end{array} \right\} \tag{9-55}$$

We have added all of the processes in which a particle emits a quasi particle of type σ and momentum $\hbar\mathbf{k}$ and subtracted all of those processes in which these quasi particles are absorbed. Substituting the transition probabilities

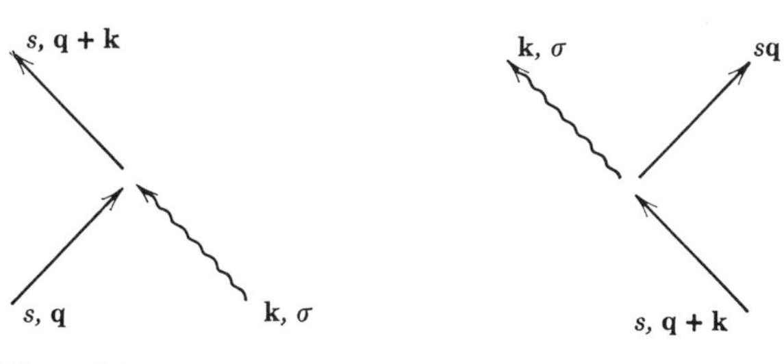

Figure 9-1

per unit time for the diagrams gives

$$\frac{\partial}{\partial t} N_\sigma(\mathbf{k}) = \sum_{\mathbf{q},s} \frac{2\pi}{\hbar}\left[\frac{4\pi e_s^2 \hbar \Omega_{\mathbf{k}\sigma}}{\Omega k^2 \left(\frac{\partial}{\partial\omega}\,\omega\varepsilon_1\right)_{\Omega_{\mathbf{k}\sigma}}}\right] \delta[E_{s\mathbf{q}+\mathbf{k}} - E_{s\mathbf{q}} - \hbar\Omega_{\mathbf{k}\sigma}]$$

$$\times \{F_{s0}(\mathbf{q}+\mathbf{k})[1 \pm F_{s0}(\mathbf{q})][N_\sigma(\mathbf{k}) + 1]$$

$$- F_{s0}(\mathbf{q})[1 \pm F_{s0}(\mathbf{q}+\mathbf{k})]N_\sigma(\mathbf{k})\}$$

$$= 2\gamma N_\sigma(\mathbf{k}) + S_\sigma(\mathbf{k}) \tag{9-56a}$$

where

$$\gamma = \sum_{\mathbf{q},s}\left[\frac{4\pi^2 e_s{}^2 \Omega_{\mathbf{k}\sigma}}{\Omega k^2\left(\frac{\partial}{\partial\omega}\,\omega\varepsilon_1\right)_{\Omega_{\mathbf{k}\sigma}}}\right][F_{s0}(\mathbf{q}+\mathbf{k}) - F_{s0}(\mathbf{q})]$$

$$\times\ \delta[E_{s,\mathbf{q}+\mathbf{k}} - E_{s\mathbf{q}} - \hbar\Omega_{\mathbf{k}\sigma}] \tag{9-56b}$$

$$S_\sigma(\mathbf{k}) = \sum_{\mathbf{q},s}\left[\frac{8\pi^2 e_s{}^2 \Omega_{\mathbf{k}\sigma}}{\Omega k^2\left(\frac{\partial}{\partial\omega}\,\omega\varepsilon_1\right)_{\Omega_{\mathbf{k}\sigma}}}\right]F_{s0}(\mathbf{q}+\mathbf{k})[1 \pm F_{s0}(\mathbf{q})]$$

$$\times\ \delta[E_{s,\mathbf{q}+\mathbf{k}} - E_{s\mathbf{q}} - \hbar\Omega_{\mathbf{k}\sigma}] \tag{9-56c}$$

We have let $F_{s0}(\mathbf{k})$ be the occupation numbers of the particle states. The plus sign is to be used if the particles are Bosons, and the minus sign is to be used if they are Fermions. The damping constant γ is seen to agree with Eq. 9-46, if the change of variable $\mathbf{p} \rightarrow \mathbf{q} + \mathbf{k}$ is made. The term $S_\sigma(\mathbf{k})$ in Eq. 9-56a is due to the spontaneous emission of quasi particles.

The plasma frequency in metals is usually sufficiently high that $\hbar\omega_{pe}$ is of the order of 10 eV. It is possible to experimentally observe discrete energy losses of high energy electrons shot through thin metal films. These discrete energy losses may be interpreted as the emission of one or more plasmons by the electron.

Problem 9-3. As particles emit and absorb quasi particles their distribution function must change. Derive an equation for the rate of change of $F_{s0}(\mathbf{p})$ by a method analogous to that of Eqs. 9-55 and 9-56. The resulting coupled equations for $N_\sigma(\mathbf{k})$ and $F_{s0}(\mathbf{p})$ are called the quasi-linear equations.

10

The Problem of Infinities in Quantum Electrodynamics

One of the most distressing features of quantum electrodynamics is that when one uses perturbation theory to calculate some quantities that are presumed to be small, they turn out in fact to be infinite. In the preceding chapters we have either circumvented or cavalierly dismissed these infinite quantities. Now we must face up to them.

ATTRACTION OF PARALLEL CONDUCTORS DUE TO QUANTUM FLUCTUATIONS OF THE FIELD

We have already encountered one of the infinities to be discussed in this chapter. This is the zero point energy

$$W = \tfrac{1}{2} \sum_{\mathbf{k}\sigma} \hbar\omega_k = \infty \tag{10-1}$$

of the electromagnetic field in a vacuum. In Chapter 2 we dismissed this infinite energy with the remark that it cancels out when energy differences are taken. It is often said that the absolute value of an energy is of no significance and an arbitrary constant can be added or subtracted. This is not always true. In general relativity the absolute value of the energy is a physically significant quantity; it determines the curvature of space.

It is not completely clear that there is a zero point energy term in the Hamiltonian. It may be that the correct Hamiltonian is

$$H_{\text{rad}} = \sum_{\mathbf{k},\sigma} \hbar\omega_k a^+_{\mathbf{k}\sigma} a_{\mathbf{k}\sigma} \tag{10-2}$$

If one works backward from this to find what the electromagnetic field energy is in terms of **E** and **B** one finds

$$H_{\text{rad}} = \frac{1}{8\pi}\int d^3x\left\{E^2 + B^2 + \frac{i}{\sqrt{-\nabla^2}}[\mathbf{E}\cdot(\nabla\times\mathbf{B}) - (\nabla\times\mathbf{B})\cdot\mathbf{E}]\right\} \quad (10\text{-}3a)$$

where the operator $1/\sqrt{-\nabla^2}$ is defined by

$$\frac{1}{\sqrt{-\nabla^2}}\,e^{i\mathbf{k}\cdot\mathbf{x}} = \frac{1}{k}\,e^{i\mathbf{k}\cdot\mathbf{x}} \quad (10\text{-}3b)$$

If the operator nature of **E** and **B** could be ignored then the last term in Eq. 10-3a vanishes and H_{rad} reduces to the classical field energy with which we began in Chapter 2. It is possible that classical theory has not been a reliable guide and that Eq. 10-3a rather than Eq. 2-12 is the correct form of the energy. If this is so then the zero point energy of the field vanishes.

There is an argument due to Casimir[53] suggesting that the zero point energy exists, and has observable consequences. This argument has been elaborated on by Lifschitz.[54] We give the argument here in its simplest form. In Chapter 2 we quantized the electromagnetic field in a cubical box of volume $\Omega = L^3$. Let us now modify this by putting conducting planes at $x = 0$ and $x = R$ as shown in Fig. 10-1. We let L become infinite but keep R finite. We denote by W_L the energy in the box when the conducting plane at $x = R$ is absent. When the conducting plane is present the energy in the box can be divided into two parts; W_R, the energy between $x = 0$ and $x = R$, and W_{L-R}, the part between $x = R$ and $x = L$. Each of these

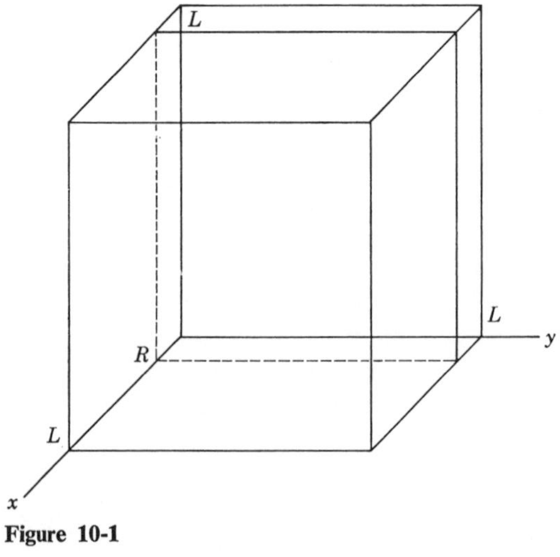

Figure 10-1

energies is divergent. Using $\omega = ck$ and letting $L \to \infty$, we find

$$W_L = \frac{\hbar c}{2} \cdot 2 \sum_{\mathbf{k}} k = \hbar c \frac{L^3}{(2\pi)^3} \iiint dk_x\, dk_y\, dk_z \sqrt{k_x^{\,2} + k_y^{\,2} + k_z^{\,2}} \tag{10-4a}$$

$$W_{L-R} = \frac{\hbar c L^2}{(2\pi)^3} (L - R) \iiint dk_x\, dk_y\, dk_z \sqrt{k_x^{\,2} + k_y^{\,2} + k_z^{\,2}} \tag{10-4b}$$

$$W_R = \frac{\hbar c L^2}{(2\pi)^2} \sum_{n=-\infty}^{+\infty} \iint dk_y\, dk_z \sqrt{k_y^{\,2} + k_z^{\,2} + \left(\frac{2\pi n}{R}\right)^2} \tag{10-4c}$$

In Eqs. 10-4a and 10-4b we have let $L \to \infty$ and replaced sums by integrals in the usual way. In Eq. 10-4c we have kept R finite so that the sum over $k_x = 2\pi n/R$ must be retained as a sum. Although each of these energies is divergent, the difference between the energy with the conducting plane at $x = R$, $W_R + W_{L-R}$, and the energy without this conducting plane, W_{L-R} is finite. Thus

$$\begin{aligned} \Delta W &= W_R + W_{L-R} - W_L \\ &= \frac{\hbar c L^2}{(2\pi)^2} \iint dk_y\, dk_z \left\{ \sum_{n=-\infty}^{+\infty} \sqrt{k_y^{\,2} + k_z^{\,2} + \left(\frac{2\pi n}{R}\right)^2} \right. \\ &\quad \left. - \frac{R}{2\pi} \int_{-\infty}^{+\infty} dk_x \sqrt{k_y^{\,2} + k_z^{\,2} + k_x^{\,2}} \right\} \end{aligned} \tag{10-5}$$

Letting

$$dk_y\, dk_z = 2\pi k_\perp\, dk_\perp \tag{10-6a}$$

$$k_z^{\,2} = k_y^{\,2} + k_z^{\,2} = \frac{4\pi^2}{R^2} x \tag{10-6b}$$

$$k_x = \frac{2\pi}{R} \omega \tag{10-6c}$$

gives

$$\Delta W = \frac{2\pi^2 \hbar c L^2}{R^3} \int_0^\infty dx \left\{ \sum_{n=-\infty}^{+\infty} \sqrt{x + n^2} - \int_{-\infty}^{+\infty} d\omega \sqrt{x + \omega^2} \right\} \tag{10-7}$$

The difference of the two infinite quantities in Eq. 10-6 can be evaluated with the result

$$\Delta W = \frac{\hbar c \pi^2 L^2}{720 R^3} \tag{10-8}$$

this gives a force per unit area of

$$F = \frac{\partial}{\partial R}\left(\frac{\Delta W}{L^2}\right) = -\frac{\hbar c \pi^2}{240 R^4} \tag{10-9}$$

It is noteworthy that this attractive force between conducting surfaces depends only on the separation R and on the universal constants $\hbar$ and c.

It does not depend on e which is a measure of the coupling of the electromagnetic field to matter. This force is of a purely quantum-electrodynamic origin. It vanishes as $\hbar \to 0$. Lifschitz[54] has extended this theory to describe the attraction of dielectric bodies and to include finite temperature effects. Experimental observations of this force have been reported.[55]

SELF ENERGY OF THE VACUUM

There are infinite corrections to the energy of the vacuum when the coupling between the radiation field and the electron-positron field is taken into account. To show this we will use the Hamiltonian of Chapter 6.

$$H = H_0 + H_I \tag{10-10a}$$

$$H_0 = \sum_{\mathbf{p},\lambda} E_{\mathbf{p},\lambda} b^+_{\mathbf{p},\lambda} b_{\mathbf{p},\lambda} + \sum_{\mathbf{k},\sigma} \hbar\omega_k a^+_{\mathbf{k}\sigma} a_{\mathbf{k}\sigma} \tag{10-10b}$$

$$H_I = -e \sum_{\mathbf{k},\sigma} \sum_{\mathbf{p},\lambda,\lambda'} \{(u^+_{\mathbf{p}+\mathbf{k},\lambda'} \boldsymbol{\alpha} \cdot \mathbf{u}_{\mathbf{k}\sigma} u_{\mathbf{p}\lambda}) b^+_{\mathbf{p}+\mathbf{k},\lambda'} b_{\mathbf{p},\lambda} a_{\mathbf{k}\sigma} + HC\} \tag{10-10c}$$

where

$$E_{\mathbf{p}\lambda} = \pm\sqrt{\hbar^2 c^2 p^2 + m^3 c^4} \tag{10-10d}$$

where the plus sign is to be taken for $\lambda = 1, 2$ and the minus sign is to be taken for $\lambda = 3, 4$. We denote the vacuum state by $|0\rangle$ and recall that it is the state with no photons, with no positive energy electrons, and with all of the negative energy states full (hence no positrons). This vacuum state is an eigenstate of H_0 with the eigenvalue

$$E_0^{(0)} = \sum_{\mathbf{p},\lambda=3,4} E_{\mathbf{p},\lambda} = -\infty \tag{10-11}$$

If we do not worry about general relativity where the absolute value of the energy is meaningful, then we can define this infinity away by defining a new zero order Hamiltonian as $H_0' = H_0 - E_0^{(0)}$. The vacuum state has the energy eigenvalue of zero for this new zero order Hamiltonian.

Next, we use perturbation theory to calculate the corrections to the energy of the vacuum due to H_I. The first order correction to $E_0^{(0)}$ is given by

$$E_0^{(1)} = \langle 0| H_I |0\rangle = 0 \tag{10-12}$$

This vanishes, since the creation and annihilation operators in H_I have no diagonal elements. The second order corrections give

$$E_0^{(2)} = \sum_I \frac{\langle 0| H_I |I\rangle\langle I |H_I |0\rangle}{E_0^{(0)} - E_I^{(0)}} \tag{10-13}$$

Now, in H_I there are terms that create an electron, create a positron (by destroying a negative energy electron), and create a photon. So we get

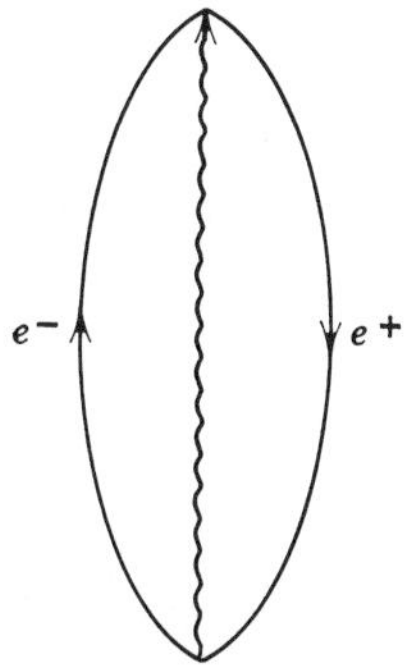

Figure 10-2

contributions to $E_0^{(2)}$ which can be represented by the diagram of Fig. 10-2. We find

$$E_0^{(2)} = \sum_{\lambda,\lambda',\sigma} \frac{\Omega^2}{(2\pi)^6} \int d^3k \int d^3p \frac{|(u^+_{\mathbf{p}+\mathbf{k},\lambda'\alpha} \cdot \mathbf{u}_{\mathbf{k}\sigma} u_{\mathbf{p}\lambda})|^2}{\hbar kc + \sqrt{\hbar^2c^2p^2 + m^2c^4} + \sqrt{\hbar^2c^2|\mathbf{p}+\mathbf{k}|^2 + m^2c^4}} \tag{10-14}$$

But this integral is clearly divergent so $E_0^{(2)} = \infty$. One would find other infinite corrections to the energy of the vacuum in higher orders of perturbation theory.

Although it is disconcerting to discover these infinities it may be argued that they are unobservable, since they always drop out whenever one calculates an observable quantity. Dirac[56] has argued that infinities of this kind, which he calls "deadwood," are of a purely mathematical nature and can be avoided if one always works in the Heisenberg representation rather than the Schrödinger representation.

RENORMALIZATION OF THE MASS OF THE ELECTRON

We now suppose that there is one electron present and consider its energy. Even in classical physics an electron of radius a has an energy of e^2/a due to the electric field that surrounds it. This energy is infinite for a point electron. In quantum theory there is an additional energy due to the transverse electromagnetic field. It is this transverse energy that is of interest to us in this section.

To do things properly we should use the relativistic theory. However, in the interest of simplicity we use the nonrelativistic interaction Hamiltonian

$$H_I = -\frac{e}{mc}\mathbf{p}\cdot\mathbf{A} + \frac{e^2}{2mc^2}A^2 \tag{10-15}$$

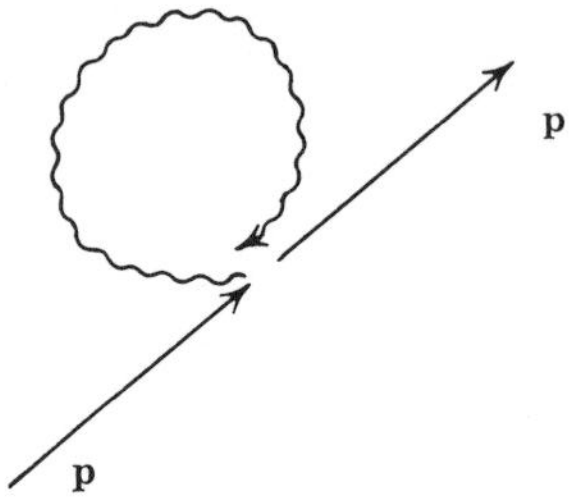

Figure 10-3

The first order correction to the energy of an electron of momentum $\hbar\mathbf{p}$ is

$$E_{\mathbf{p}}^{(1)} = \frac{e^2}{2mc^2} \langle \mathbf{p} | A^2 | \mathbf{p} \rangle \tag{10-16}$$

where $|\mathbf{p}\rangle$ denotes a state with one electron of momentum $\hbar\mathbf{p}$ and no photons. There is no contribution from the $\mathbf{p} \cdot \mathbf{A}$ term in first order, since this term connects states that differ by one photon. The A^2 term contains operators that can create and destroy the same photon. Equation 10-16 can be represented by the diagram of Fig. 10-3. Using Eq. 3-5b we find

$$E_{\mathbf{p}}^{(1)} = \frac{e^2}{2mc^2} \sum_{\mathbf{k}\sigma} \left(\frac{2\pi\hbar c^2}{\Omega} \right) \frac{\mathbf{u}_{\mathbf{k}\sigma} \cdot \mathbf{u}_{\mathbf{k}\sigma}}{kc} = \frac{2}{\pi} \frac{e^2\hbar}{mc} \int_0^\infty k\, dk = \infty \tag{10-17}$$

This gives an infinite contribution to the energy of a free electron. It is independent of $\mathbf{p}$, so it is the same for all electrons. It will cancel out whenever energy differences occur.

A more interesting infinite energy comes from the $\mathbf{p} \cdot \mathbf{A}$ term in second order. This is

$$E_{\mathbf{p}}^{(2)} = \frac{e^2}{m^2c^2} \sum_I \frac{\langle \mathbf{p} | \mathbf{p} \cdot \mathbf{A} | I \rangle \langle I | \mathbf{p} \cdot \mathbf{A} | \mathbf{p} \rangle}{E_{\mathbf{p}} - E_I} \tag{10-18}$$

Since $\mathbf{A}$ is linear in the creation and annihilation operators for photons, the intermediate states must contain one photon. The terms in Eq. 10-18 can be

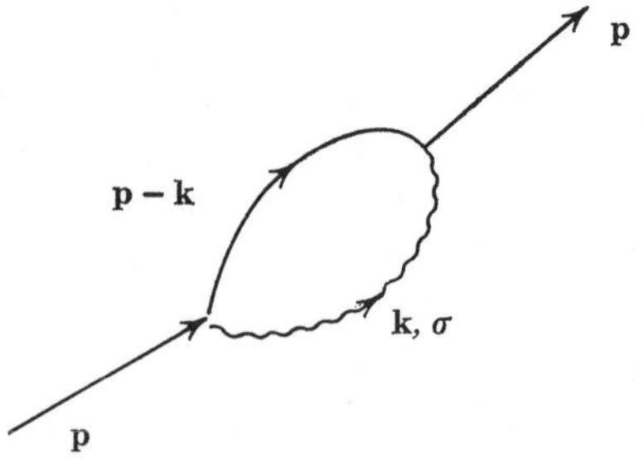

Figure 10-4

represented by the diagram of Fig. 10-4. Using Eq. 3-5a we find

$$E_{\mathbf{p}}^{(2)} = \frac{e^2}{m^2c^2} \sum_{\mathbf{k}\sigma} \left(\frac{2\pi\hbar c^2}{\Omega kc}\right) \frac{|\hbar \mathbf{p} \cdot \mathbf{u}_{\mathbf{k}\sigma}|^2}{\hbar^2 p^2/2m - (\hbar^2/2m\,|\mathbf{p} - \mathbf{k}|^2 + hkc)} \tag{10-19}$$

In evaluating this we make the dipole approximation $k \ll p$. Also, we use

$$\sum_{\sigma} |\mathbf{p} \cdot \mathbf{u}_{\mathbf{k}\sigma}|^2 = \mathbf{p} \cdot \left(\sum_{\sigma} \mathbf{u}_{\mathbf{k}\sigma}\mathbf{u}_{\mathbf{k}\sigma}\right) \cdot \mathbf{p} = \mathbf{p} \cdot \left(1 - \frac{\mathbf{k}\mathbf{k}}{k^2}\right) \cdot \mathbf{p} = p^2(1 - \cos^2\theta) \tag{10-20}$$

We obtain

$$\begin{aligned} E_{\mathbf{p}}^{(2)} &= -\frac{e^2}{m^2c^2}\left(\frac{2\pi\hbar c^2}{\Omega c}\right)\frac{\Omega}{(2\pi)^3}\int \frac{d^3k}{k}\frac{(\hbar p)^2(1 - \cos^2\theta)}{\hbar kc} \\ &= -\frac{2}{3\pi}\frac{e^2}{m^2c^2}(\hbar p)^2\int_0^\infty dk = \infty \end{aligned} \tag{10-21}$$

Again we find an infinite result, but now $E_{\mathbf{p}}^{(2)}$ is proportional to p^2. We can combine it with the zero order energy to write

$$\begin{aligned} E_{\mathbf{p}} &\simeq E_p^{(0)} + E_{\mathbf{p}}^{(2)} \\ &= \frac{1}{2m}(\hbar p)^2\left(1 - \frac{4}{3\pi}\frac{e^2}{mc^2}\int_0^\infty dk\right) \\ &= \frac{1}{2m_{\text{exp}}}(\hbar p^2) \end{aligned} \tag{10-22}$$

Now, we can adopt the following point of view. The mass m in the formula $E_{\mathbf{p}}^{(0)} = \hbar^2p^2/2m$ is the mass of a "bare" electron which does not interact with the electromagnetic field. It is fictitious, since the interaction cannot be turned off. The experimental mass of the electron must include the ever-present interaction with the field. To lowest order in the interaction it is given by

$$\frac{1}{m_{\text{exp}}} = \frac{1}{m}\left(1 - \frac{4}{3\pi}\frac{e^2}{mc^2}\int_0^\infty dk\right) \tag{10-23}$$

which must be a finite quantity whatever the fictitious mass may be. This shift of the mass of the electron from its bare value of m to its observed value of m_{exp} is called "renormalization" of the mass.

THE LAMB SHIFT

According to the Dirac theory, the $2s$ and $2p_{1/2}$ levels of the hydrogen atom should coincide. However, in some very beautiful experiments Lamb and Retherford[57] showed that there was a small energy difference between

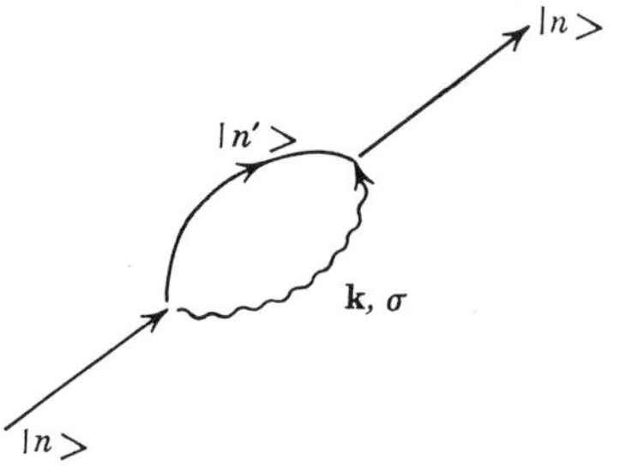

Figure 10-5

these levels corresponding to a frequency of about 1000 megacycles. This is known as the Lamb shift. It was suspected that this shift was due to the interaction with the electromagnetic field, but when calculations were made of the shift it turned out to be infinite. Bethe[5] showed that this difficulty could be overcome by the renormalization of the mass.

We shall follow Bethe in making a nonrelativistic calculation. The terms responsible for the shift in the energy of the state $|n\rangle$ of the hydrogen atom can be represented by the diagram of Fig. 10-5. This diagram is analogous to Fig. 10-4, the only difference being that the states of the electron are bound states rather than free states. The correction to the energy is

$$E_n^{(2)} = \frac{e^2}{m^2c^2} \sum_{n'} \sum_{\mathbf{k},\sigma} \left(\frac{2\pi\hbar c^2}{\Omega k c}\right) \frac{|\langle n'|\ \mathbf{p}_0 \cdot \mathbf{u}_{\mathbf{k}\sigma}\ |n\rangle|^2}{E_n - E_{n'} - \hbar k c} \tag{10-24}$$

(We have used $\mathbf{p}_0 = (\hbar/i)\ \partial/\partial\mathbf{x}$ to avoid confusion with the wave vector $\mathbf{p}$ which was used in the preceding section. We have made the dipole approximation in Eq. 10-24.) Using Eq. 10-20 we can simplify Eq. 10-24 to obtain

$$E_n^{(2)} = \frac{2}{3\pi} \frac{e^2}{m^2c} \sum_{n'} \int_0^\infty k\, dk \frac{|\langle n'|\ p_0\ |n\rangle|^2}{E_n - E_{n'} - hkc} \tag{10-25}$$

Equation 10-25 is divergent just as Eq. 10-19 was. However, Bethe reasoned as follows. For a free electron $\mathbf{p}_0$ has only diagonal matrix elements and Eq. 10-25 reduces to Eq. 10-21 which we interpreted as the change in the kinetic energy due to the fact that electromagnetic mass is added to the mass of the electron. For a bound electron the square of the momentum in Eq. 10-21 should be replaced by its expectation value $\langle n|\ p_0^2\ |n\rangle$. By the completeness relation

$$\langle n|\ p_0^2\ |n\rangle = \sum_{n'} |\langle n'|\ p_0\ |n\rangle|^2 \tag{10-26}$$

Thus the correction to the kinetic energy due to electromagnetic mass is

$$-\frac{2}{3\pi} \frac{e^2}{m^2c} \sum_{n'} \int_0^\infty k\, dk \frac{|\langle n'|\ p_0\ |n\rangle|^2}{hkc} \tag{10-27}$$

This should be subtracted from Eq. 10-25 to obtain the observable energy level shift of

$$\Delta E_n^{(2)} = \frac{2}{3\pi}\frac{e^2}{m^2c}\sum_{n'}\int_0^\infty k\,dk\,|\langle n'|\,p_0\,|n\rangle|^2\left[\frac{1}{E_n - E_{n'} - hkc} + \frac{1}{hkc}\right] \tag{10-28}$$

This integral is still divergent, but now only logarithmically. Bethe reasoned (his reasons will be discussed presently) that in a relativistic theory the integral should be convergent. This convergence can be simulated in the nonrelativistic theory by cutting off the integral when the energy of the photon $\hbar ck$ becomes comparable to mc^2, the rest energy of the electron. Replacing the upper limit of the integral by $mc/\hbar$ and carrying out the integration gives

$$\Delta E_n^{(2)} = \frac{2}{3\pi}\frac{e^2}{\hbar c^3m^2}\sum_{n'}|\langle n'|\,p\,|n\rangle|^2\,(E_{n'} - E_n)\cdot\log\left|\frac{mc^2}{E_{n'} - E_n}\right| \tag{10-29}$$

where $E_{n'} - E_n$ has been neglected compared with mc^2. In evaluating Eq. 10-24 it is a good approximation to replace $(E_{n'} - E_n)$ in the argument of the logarithm by an appropriately chosen average value since the argument is large and the logarithm is a slowly varying function. Then the logarithm may be removed from the sum. The sum that remains can be evaluated as follows.

$$\begin{aligned}\sum_{n'}|\langle n'|\,\mathbf{p}\,|n\rangle|^2\,(E_{n'} - E_n) &= \sum_{n'}\langle n|\,\mathbf{p}\,|n'\rangle\langle n'|\,\mathbf{p}\,|n\rangle(E_{n'} - E_n)\\ &= \sum_{n'}\langle n|\,\mathbf{p}(H_0 - E_n)\,|n'\rangle\langle n'|\,\mathbf{p}\,|n\rangle\\ &= \langle n|\,\mathbf{p}(H_0 - E_n)\cdot\mathbf{p}\,|n\rangle\\ &= \langle n|\,\mathbf{p}\cdot(H_0\mathbf{p} - \mathbf{p}H_0)\,|n\rangle\end{aligned} \tag{10-30}$$

where H_0 is the Hamiltonian of the hydrogen atom. We may use

$$H_0\mathbf{p} - \mathbf{p}H_0 = -\frac{\hbar}{i}\frac{\partial}{\partial\mathbf{x}}V \tag{10-31}$$

where $V = Ze^2/r$ to write Eq. 10-30 as

$$\begin{aligned}\langle n|\,\mathbf{p}\cdot(H_0\mathbf{p} - \mathbf{p}H_0)\,|n\rangle &= \hbar^2\int\psi_n^*\nabla\cdot(\nabla V\psi_n)\,d^3x\\ &= -\tfrac{1}{2}\hbar^2\int|\psi_n|^2\,\nabla^2V\,d^3x\\ &= +\frac{\hbar^2}{2}\,4\pi Ze^2\int|\psi_n|^2\,\delta(\mathbf{r})\,d^3x\\ &= +2\pi Ze^2\hbar^2\,|\psi_n(0)|^2\end{aligned} \tag{10-32}$$

Finally we get

$$\Delta E_n^{(2)} = \tfrac{4}{3}Z\frac{e^4\hbar}{m^2c^3}\,|\psi_n(0)|^2\log\left|\frac{mc^2}{(E_{n'} - E_n)_{av}}\right| \tag{10-33}$$

The $\psi_n(0)$ vanishes for states with nonzero angular momentum. For s states

$$|\psi_n(0)|^2 = \frac{1}{\pi}\left(\frac{Z}{na}\right)^3 \tag{10-34}$$

where n is the principal quantum number and a is the Bohr radius.

Bethe calculated numerically $(E_{n'} - E_n)_{\text{av}}$. When he applied his results to the Lamb shift he found a value of 1040 megacycles which is in good agreement (considering the approximations made) with the experimental value of 1057 megacycles.

The problem of the self energy of the electron is not as bad in the relativistic theory as it is in the nonrelativistic theory for the following reason. If one tries to construct a wave packet to represent a positive energy electron using relativistic wave functions then the negative energy states are not available for use, since they are already full. As a result there is a limit to how small the wave packet can be made. It is as if the electron had a finite size about equal to its Compton wavelength. It turns out that this does not completely cure the divergence but makes it only logarithmically divergent instead of the linear divergence of Eq. 10-21. Bethe argued that when the subtraction was made in the relativistic theory to obtain the equivalent of our Eq. 10-28 a convergent result would be obtained. When the relativistic calculation was done by Kroll and Lamb[59] this was found to be the case. The agreement between experiment and the relativistic theory is now extremely good.

It is now realized that all of the infinities in quantum electrodynamics are essentially unobservable, since they must be embodied in the finite values of the observed mass and charge of the particles. This favorable circumstance has made quantum electrodynamics an extremely successful theory in spite of the infinities that detract from its esthetic appeal.

ANOMALOUS MAGNETIC MOMENT OF THE ELECTRON

The Dirac equation predicts a magnetic moment of the electron of $-e\hbar/2mc$. When the coupling of the electromagnetic field is taken into account there is a shift in this value. We give a very simplified theory of this effect and then qualitatively discuss the more rigorous theory.

Consider an electron fixed at the origin of the coordinate system. Let there be a uniform magnetic field B in the z-direction. The zeroth order Hamiltonian is

$$H_0 = -\boldsymbol{\mu} \cdot \mathbf{B} = \frac{e\hbar B}{2mc}\sigma_z = g\frac{\hbar\omega_0}{2}\sigma_z \tag{10-35}$$

where $g = 2$ and $\omega_0 = eB/2mc$. The eigenvalues of H_0 are

$$E_m^{(0)} = mg\hbar\omega \tag{10-36a}$$

where

$$m = \pm\tfrac{1}{2} \tag{10-36b}$$

Associated with the radiation field there will be a magnetic field given by

$$\nabla \times \mathbf{A} = i \sum_{\mathbf{k}\sigma} \left(\frac{2\pi\hbar c^2}{\Omega\omega_k}\right)^{1/2} (\mathbf{k} \times \mathbf{u}_{\mathbf{k}\sigma})[a_{\mathbf{k}\sigma} e^{i\mathbf{k}\cdot\mathbf{x}} - a_{\mathbf{k}\sigma}^{+} e^{-i\mathbf{k}\cdot\mathbf{x}}] \tag{10-37}$$

This also interacts with the magnetic moment of the electron to give the energy

$$\begin{aligned} H_I &= -\boldsymbol{\mu} \cdot (\nabla \times \mathbf{A}) \\ &= -\frac{ie\hbar}{2mc} \sum_{\mathbf{k}\sigma} \left(\frac{2\pi\hbar c^2}{\Omega\omega_k}\right)^{1/2} [\boldsymbol{\sigma} \cdot (\mathbf{k} \times \mathbf{u}_{\mathbf{k}\sigma})][a_{\mathbf{k}\sigma} e^{i\mathbf{k}\cdot\mathbf{x}} - a_{\mathbf{k}\sigma}^{+} e^{-i\mathbf{k}\cdot\mathbf{x}}] \end{aligned} \tag{10-38}$$

(The σ when set in boldface or with subscripts x, y, or z denotes Pauli matrices; there should be no confusion with the subscript σ denoting polarization.)

Just as in the preceding section, H_1 can give a second order correction to the energy of

$$E_m^{(2)} = \sum_I \frac{\langle m|\, H_I \,|I\rangle\langle I|\, H_I \,|m\rangle}{E_m^{(0)} - E_I} \tag{10-39}$$

This correction can be represented by the diagram of Fig. 10-6. We have denoted by $|m\rangle$ the state in which the electron has the quantum number m and there are no photons. In the intermediate state the electron has changed its quantum number to m' and there is a photon present. By the artifice of fixing the electron at the origin we have restricted the electron to two possible states. This simplifies the problem but at the expense of making it somewhat

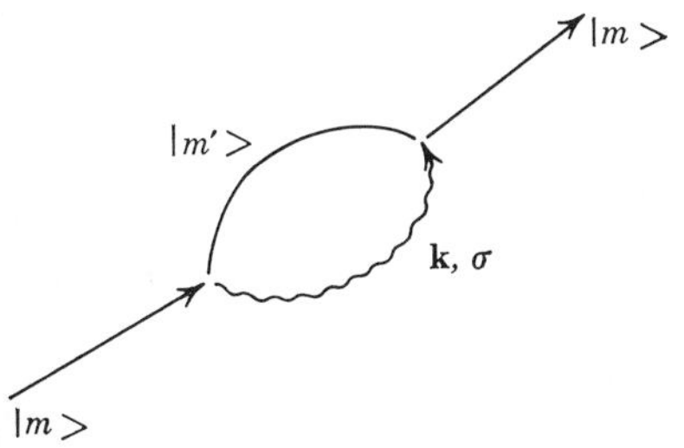

Figure 10-6

artificial. Using Eq. 10-38 in Eq. 10-39 gives

$$E_m^{(2)} = \sum_{m'}\sum_{\mathbf{k}\sigma} \frac{e^2\hbar^2}{4m^2c^2}\left(\frac{2\pi\hbar c^2}{\Omega\omega_k}\right)\frac{|\langle m'|\,\boldsymbol{\sigma}\cdot\mathbf{k}\times\mathbf{u}_{\mathbf{k}\sigma}\,|m\rangle|^2}{g\hbar\omega_0(m-m')-\hbar kc}$$

$$= \frac{e^2\hbar^3}{16\pi^2m^2c}\sum_{m'}\sum_{\sigma}\int\frac{d^3k}{k}\frac{|\langle m|\,\boldsymbol{\sigma}\cdot\mathbf{k}\times\mathbf{u}_{\mathbf{k}\sigma}\,|m\rangle|^2}{g\hbar\omega_0(m-m')-\hbar kc} \tag{10-40}$$

We can carry out the sum over polarizations and show that

$$\sum_{\sigma}|\langle m'|\,\boldsymbol{\sigma}\cdot\mathbf{k}\times\mathbf{u}_{\mathbf{k}\sigma}\,|m\rangle|^2 = k^2\,|\langle m'|\,\boldsymbol{\sigma}\,|m\rangle|^2 - |\langle m'|\,\mathbf{k}\cdot\boldsymbol{\sigma}\,|m\rangle|^2 \tag{10-41}$$

If we choose $m = +\frac{1}{2}$ then this is

$$k^2(1-\cos^2\theta) \tag{10-42a}$$

for $m' = +\frac{1}{2}$ and

$$k^2(1+\cos^2\theta) \tag{10-42b}$$

for $m' = -\frac{1}{2}$. Here $\cos\theta = k_z/k$. The angular integrations can be carried out with the result

$$E_m^{(2)} = \frac{e^2\hbar^2}{6\pi m^2c^2}\int_0^\infty k^3\,dk\left\{-\frac{1}{k}+\frac{2}{g\omega_0/c-k}\right\} \tag{10-43}$$

This is infinite, of course. Furthermore, it is infinite even when B (which is contained in ω_0) vanishes. We should subtract this energy which remains when $B = 0$, for it cannot be interpreted as an energy of interaction of a magnetic moment with a magnetic field. Subtracting a term with the brace replaced by $-3/k$ and calling the difference $\Delta E_m^{(2)}$ we obtain

$$\Delta E_m^{(2)} = \frac{e^2\hbar^2 g\omega_0}{3\pi m^2c^3}\int_0^\infty\frac{k^2\,dk}{g\omega_0/c-k} \tag{10-44}$$

The integral is still divergent, so we shall cut off the integral at the upper limit of $k = mc/\hbar$ as we did in the Lamb shift calculation. The result is

$$\Delta E_m^{(2)} = -g\frac{\hbar\omega_0}{2}\left(\frac{e^2}{3\pi hc}\right)\left\{1+\left(\frac{2g\hbar\omega_0}{mc^2}\right)+\left(\frac{2g\hbar\omega_0}{mc^2}\right)^2\log\left|\frac{mc^2-g\hbar\omega_0}{gh\omega_0}\right|\right\} \tag{10-45}$$

If we keep only the term that is linear in B we obtain

$$\Delta E_m^{(2)} = -g\frac{\hbar\omega_0}{2}\left(\frac{e^2}{3\pi hc}\right) \tag{10-46}$$

A correct relativistic treatment gives the factor in Eq. 10-46 as $(e^2/2\pi\hbar c)$ rather than $(e^2/3\pi\hbar c)$. Our crude approximate treatment has led to a result that is only off by a factor of $\frac{2}{3}$. We could have gotten the right answer if we had chosen the cut off to be $\sqrt{\frac{3}{2}}\,(mc/\hbar)$. The answer is much more sensitive to the cutoff than it was in the case of the Lamb shift.

The terms in Eq. 10-45 which are nonlinear in B can be interpreted as a magnetic polarizability of the electron. They are too small to be measurable.

We conclude by describing how a correct calculation would go. First, one must use the solutions of the Dirac equation in a uniform magnetic field

$$\left[c\boldsymbol{\alpha}\cdot\left(\frac{\hbar}{i}\nabla - \frac{e}{c}\mathbf{A}_0\right) + \beta mc^2\right]\psi_n = E_n\psi_n \tag{10-47}$$

These solutions are well known.[61] (The n stands for the four quantum numbers that characterize a solution.) In the usual way we expand

$$\psi = \sum_n b_n\psi_n$$

and find for the unperturbed Hamiltonian

$$\begin{aligned} H_0 &= \int d^3x\psi^+\left[\bar{c}\boldsymbol{\alpha}\cdot\left(\frac{\hbar}{i}\nabla - \frac{e}{c\mathbf{A}_0}\right) + \beta mc^2\right]\psi + \frac{1}{8\pi}\int d^3x(E^2 + B^2) \\ &= \sum_n E_n b_n^+ b_n + \sum_{\mathbf{k}\sigma} \hbar\omega_k a_{\mathbf{k}\sigma}^+ a_{\mathbf{k}\sigma} \end{aligned} \tag{10-48}$$

The interaction with the radiation field gives the term

$$\begin{aligned} H_I &= -e\int d^3x\psi^+\boldsymbol{\alpha}\cdot\mathbf{A}\psi \\ &= -e\sum_{\mathbf{k},\sigma}\sum_{n,n'}\{b_n^+ b_{n'} a_{\mathbf{k}\sigma}\int d^3x\psi_n^*\boldsymbol{\alpha}\cdot\mathbf{u}_{\mathbf{k}\sigma}\psi_{n'}e^{i\mathbf{k}\cdot\mathbf{x}} + \text{HC}\} \end{aligned} \tag{10-49}$$

The electrostatic interaction cannot be neglected. The methods of Chapter 8 may be used to write it as

$$\begin{aligned} H_c &= \frac{e^2}{2}\int d^3x\, d^3x'\, \frac{\psi^+(\mathbf{x})\psi^+(\mathbf{x}')\psi(\mathbf{x})\psi(\mathbf{x}')}{|\mathbf{x} - \mathbf{x}'|} \\ &= \frac{e^2}{2}\sum_{nn'll'} a_n^+ a_{n'} a_l^+ a_{l'}\int d^3x\, d^3x'\, \frac{\psi_n^+(\mathbf{x})\psi_{n'}(\mathbf{x})\psi_l^+(\mathbf{x}')\psi_{l'}(\mathbf{x}')}{|\mathbf{x} - \mathbf{x}'|} \end{aligned} \tag{10-50}$$

It might be thought that since only one electron is involved the Coulomb interaction would not enter. However, it must be kept in mind that in relativistic quantum electrodynamics there is no such thing as a one body problem. There is always present the infinite sea of negative energy electrons.

One chooses a state of H_0 in which there is one electron in state n and no photons. Perturbation theory is used to calculate the shift in the energy of this state. The H_c gives a correction to the energy proportional to e^2 in first order and H_I gives an e^2 correction in second order. When the part of the energy that remains when $B = 0$ is subtracted, both contributions are convergent. The H_c contributes

$$\Delta E_n^{(1)} = +g\frac{\hbar\omega_0}{2}\left(\frac{e^2}{3\pi\hbar c}\right) \tag{10-51}$$

and H_I contributes

$$\Delta E_n^{(2)} = -g\frac{\hbar\omega_0}{2}\left(\frac{5e^2}{6\pi\hbar c}\right) \tag{10-52}$$

so that the net result is

$$\Delta E = -g\frac{\hbar\omega_0}{2}\left(\frac{e^2}{2\pi\hbar c}\right) \tag{10-53}$$

This correction can be considered to be a correction to the g-factor of

$$\Delta g = \frac{e^2}{2\pi\hbar c} = 0.001161 \tag{10-54}$$

Further contributions to the magnetic moment of the electron arise from corrections of order e^4, e^6 and so on. The e^4 correction has been worked out with the result

$$\Delta g = \frac{e^2}{2\pi\hbar c} - \frac{2.97}{\pi^2}\left(\frac{e^2}{\hbar c}\right)^2 = 0.001145 \tag{10-55}$$

This result is in excellent agreement with experiment.

Problem 10-1. A calculation of the Lamb shift due to Welton is very instructive in that it makes it clear that the origin of the shift is the zero point fluctuations of the electromagnetic field. It proceeds as follows. Solve the classical equations of motion for a particle in an oscillating electric field. Calculate the mean square displacement of the electron assuming that it is acted on by a superposition of electromagnetic waves and assigning to each mode of this radiation field the energy $\hbar\omega/2$. Show that the mean square displacement is

$$\langle|\mathbf{\Delta x}|^2\rangle = \frac{2}{\pi}\frac{e^2}{\hbar^2}\left(\frac{\hbar}{mc}\right)^2\int_0^\infty \frac{dk}{k} \tag{10-56}$$

This is infinite, of course. It must be made finite by a suitable choice of cutoffs at the upper and lower limits of integration. Next, show that the change in potential energy of the electron due to fluctuations in the electrons position is

$$V(\mathbf{x} + \mathbf{\Delta x}) = [1 + \mathbf{\Delta x}\cdot\nabla + \tfrac{1}{2}(\mathbf{\Delta x}\cdot\nabla)^2 + \cdots]V(\mathbf{x}) \tag{10-57}$$

and

$$\langle V(\mathbf{x} + \mathbf{\Delta x})\rangle = \left[1 + \frac{1}{6}\langle|\mathbf{\Delta x}|^2\rangle\nabla^2\right]V(\mathbf{x}) \tag{10-58}$$

Treat this second term as a perturbation and use perturbation theory to show that the shift in energy of a level in the hydrogen atom is

$$\Delta E = \frac{4}{3}\frac{e^2}{\hbar c}\left(\frac{\hbar}{mc}\right)^2 \log\frac{k_{\max}}{k_{\min}}|\psi(0)|^2$$

Appendix A

Relativistic Wave Equations

In discussions of relativistic invariance two notations are in common use. In one of these we denote the space-time coordinates by

$$x = x_1, \quad y = x_2, \quad z = x_3, \quad ict = x_4 \tag{A-1}$$

In this notation the interval between two neighboring events is

$$(ds)^2 = dx_\mu\, dx_\mu = (dx_1)^2 + (dy)^2 + (dz)^2 - c^2t^2 \tag{A-2}$$

In the other notation

$$x_0 = ct, \quad x_1 = x, \quad x_2 = y, \quad x_3 = z \tag{A-3}$$

and the interval between neighboring events is

$$(ds)^2 = g_{\lambda\mu}\, dx_\lambda\, dx_\mu = -c^2t^2 + (dx)^2 + (dy)^2 + (dz)^2 \tag{A-4}$$

where $g_{\lambda\mu} = 0, \quad \lambda \neq \mu$

$g_{00} = -1$

$g_{11} = g_{22} = g_{33} = 1$

We are using the summation convention; a repeated index is to be summed over. For simplicity we refer to these notations as the x_4 notation and the x_0 notation.

Under a Lorentz transformation the coordinates transform as

$$x'_\mu = a_{\mu\lambda} x_\lambda \tag{A-5}$$

where the elements of $a_{\mu\lambda}$ are restricted by the requirement that $(ds)^2$ be an invariant. For the x_4 notation this requirement is $a_{\mu\lambda}a_{\mu\lambda} = \delta_{\lambda\gamma}$; for the x_0-notation the requirement is $g_{\lambda\mu}a_{\lambda\alpha}a_{\mu\beta} = g_{\alpha\beta}$.

It is a postulate of the special theory of relativity that physical laws are invariant under Lorentz transformations; that is, they take the same form in a reference system with space-time coordinates x'_μ as they do in a system with space-time coordinates x_μ. This is easily accomplished if the physical laws are written as relations among tensors of the same rank, since then all of the terms in the equation transform in the same way under the Lorentz transformation, Eq. A-5. This postulate of relativistic invariance is one of the guiding principles to be followed in constructing relativistic wave equations.

The other guiding principle is that the frequency of the wave must be related to the energy of the corresponding particle by the Einstein relation, and the wave vector of the wave must be related to the momentum of the corresponding particle by the De Broglie relation; that is

$$\begin{aligned} E &= \hbar\omega \qquad \text{(Einstein)} \\ \mathbf{p} &= \hbar\mathbf{k} \qquad \text{(De Broglie)} \end{aligned} \tag{A-6}$$

If one constructs a wave packet of the form

$$\psi(\mathbf{x}, t) = \int d^3k C(\mathbf{k}) e^{i(\mathbf{k}\cdot\mathbf{x} - \omega(\mathbf{k})t)} \tag{A-7}$$

and uses $E = E(\mathbf{p})$ to infer the functional form of $\omega = \omega(\mathbf{k})$, then it follows that the centroid of the wave packet moves with the velocity of the corresponding classical particle. This follows from

$$\frac{\partial E(\mathbf{p})}{\partial \mathbf{p}} = \frac{\partial \omega(\mathbf{k})}{\partial \mathbf{k}} \tag{A-8}$$

The left-hand side of Eq. A-8 is the particle velocity given by Hamilton's equations and the right-hand side is the group velocity of a wave packet. The equation satisfied by $\psi(\mathbf{x}, t)$ may be obtained by writing

$$[E - E(\mathbf{p})]\psi(\mathbf{x}, t) = 0 \tag{A-9}$$

and replacing E by $-\hbar/i(\partial/\partial t)$ and $\mathbf{p}$ by $\hbar/i(\partial/\partial\mathbf{x})$, for then

$$\begin{aligned} &\left[-\frac{\hbar}{i}\frac{\partial}{\partial t} - E\left(\frac{\hbar}{i}\frac{\partial}{\partial \mathbf{x}}\right)\right]\psi(\mathbf{x}, t) \\ &\qquad = \int d^3k (C(\mathbf{k})[h\omega(\mathbf{k}) - E(\hbar\mathbf{k})] e^{i(\mathbf{k}\cdot\mathbf{x} - \omega(\mathbf{k})t)} = 0 \end{aligned} \tag{A-10}$$

If one chooses the nonrelativistic relation between energy and momentum

$$E = \frac{1}{2m} p^2 \tag{A-11}$$

then one obtains the nonrelativistic Schrödinger equation

$$\frac{\hbar}{i}\frac{\partial\psi}{\partial t} = \frac{\hbar^2}{2m}\nabla^2\psi \tag{A-12}$$

If one chooses the relativistic relation

$$E = \sqrt{c^2p^2 + m^2c^4} \tag{A-13}$$

one obtains

$$\frac{h}{i}\frac{\partial\psi}{\partial t} = \sqrt{-\hbar^2c^2\nabla^2 + m^2c^4}\ \psi \tag{A-14}$$

which is rather troublesome to interpret. Actually, the right-hand side can be interpreted as a nonlocal operator, but the prevailing view of the theorist is that nonlocal operators should be avoided if at all possible. A better solution is to square both sides of Eq. A-13 and then replace E and $\mathbf{p}$ by the corresponding operators to obtain

$$-\hbar^2\frac{\partial^2\psi}{\partial t^2} = -\hbar^2c^2\nabla^2\psi + m^2c^4\psi \tag{A-15}$$

which may also be written

$$\Box^2\psi = \left(\frac{mc}{\hbar}\right)^2\psi \tag{A-16}$$

where

$$\Box^2 = \nabla^2 - \frac{1}{c^2}\frac{\partial^2}{\partial t^2} = \frac{\partial^2}{\partial x_\mu\,\partial x_\mu} \quad \text{or} \quad g_{\lambda\mu}\frac{\partial^2}{\partial x_\lambda\,\partial x_\mu} \tag{A-17}$$

is called the *D*'Alembentian operator. It is easily shown that $\Box^2$ is a scalar operator.

Problem A-1. Use $a_{\mu\lambda}a_{\mu\gamma} = \delta_{\lambda\gamma}$ (x_4 notation) to show that the transformation inverse to Eq. A-5 is

$$x = a_{\mu\lambda}x'_\mu \tag{A-18}$$

Use $x_\lambda x_\lambda = x'_\mu x'_\mu$ to show that $a_{\mu\lambda}a_{\beta\lambda} = \delta_{\mu\beta}$. Show that

$$\frac{\partial}{\partial x_\mu} = a_{\lambda\mu}\frac{\partial}{\partial x'_\lambda} \tag{A-19}$$

and that

$$\frac{\partial^2}{\partial x'_\mu\,\partial x'_\mu} = \frac{\partial^2}{\partial x_\lambda\,\partial x_\lambda} \tag{A-20}$$

If $\psi(x_\mu)$ is a scalar then Eq. A-16 is a relativistically invariant equation; it is called the Klein-Gordon equation, although it was originally proposed by Schrödinger in 1926. Equation A-16 will be relativistically invariant if $\psi(x_\mu)$ is a tensor of any rank. If ψ is a 4-vector, we write it as $\psi_\nu(x_\mu)$ and the resulting equation is called the Proca equation. In the special case that the rest mass of the particle vanishes, the equation

$$\Box^2 \psi_\nu = 0 \tag{A-21}$$

is just the equation obeyed by the vector potential for the electromagnetic field.

We could go on to consider second rank tensors $\psi_{\mu\nu}$, third rank tensors $\psi_{\mu\nu\lambda}$, and so on. The subscripts on ψ may be considered as additional coordinates of the particle. For instance, the particle described by the 4-vector wave function $\psi_\nu(x_\mu)$ has, in addition to the space-time coordinates x_μ, the coordinate ν which can only take on the discrete values $\nu = 1, 2, 3, 4$. It may be shown that these additional coordinates are related to the spin of the particle. A scalar wave function describes a spin-zero particle; a 4-vector wave function describes a spin-one particle; a second rank tensor wave function describes a spin-two particle; and so on.

As a single particle equation, Eq. A-16 has some undesirable features which are connected with the fact that it is a second order equation in the time. As a result it was in disrepute for about seven years after it was proposed. Then in 1934 Pauli and Weiskopf reestablished the validity of the equation by reinterpreting it as a field equation which was to be quantized as the electromagnetic field equations were. It is now believed to be the equation that describes mesons.

It may appear that by considering tensors of all ranks as choices for ψ we have exhausted all of the possibilities for relativistically invariant wave equations. However, as Edington phrased it, "something has slipped through the net." Dirac reasoned that if a relativistic equation is to be first order in time then in order for space and time to be treated symmetrically it must be first order in space as well. Let us try to extract the square root in Eq. A-13 by writing

$$E = c(\alpha_1 p_x + \alpha_2 p_y + \alpha_3 p_z) + \beta mc^2 \tag{A-22}$$

Replacing E by $-\hbar/i\ \partial/\partial t$ and p_i by $\hbar/i\ \partial/\partial x_i$ and letting each side operate on ψ gives an equation which is first order in both space and time derivatives. If the right-hand side of Eq. A-22 is indeed the square root of $c^2p^2 + m^2c^4$ then we must have

$$\begin{aligned} c^2p^2 + m^2c^4 = {} & c^2(\alpha_1^2 p_x^2 + \alpha_2^2 p_y^2 + \alpha_3^2 p_z^2) \\ & + \beta^2 m^2 c^4 + 2c^2(\alpha_1\alpha_2 + \alpha_2\alpha_1)p_x p_y + \cdots \end{aligned} \tag{A-23}$$

This cannot be accomplished if α_1, α_2, α_3, and β are numbers, but it is possible if they are noncommuting matrices which satisfy

$$\begin{aligned} \alpha_1^2 = \alpha_2^2 = \alpha_3^2 = \beta^2 &= 1 \\ \alpha_i\alpha_j + \alpha_j\alpha_i &= 0 \qquad \text{for } i \neq j \\ \alpha_i\beta + \beta\alpha_i &= 0 \qquad \text{for all } i \end{aligned} \tag{A-24}$$

It can be shown that the lowest order matrices which satisfy these relations are 4 × 4. A convenient choice for the 4 × 4 matrices is

$$\boldsymbol{\alpha} = \begin{pmatrix} 0 & \boldsymbol{\sigma} \\ \boldsymbol{\sigma} & 0 \end{pmatrix} \qquad \beta = \begin{pmatrix} 1 & 0 \\ 0 & -1 \end{pmatrix} \tag{A-25}$$

where $\boldsymbol{\alpha} = (\alpha_1, \alpha_2, \alpha_3)$, and we have written the 4 × 4 matrices as 2 × 2 matrices of 2 × 2 matrices. We have denoted by $\boldsymbol{\sigma} = (\sigma_1, \sigma_2, \sigma_3)$ the Pauli matrices

$$\sigma_1 = \begin{pmatrix} 0 & 1 \\ 1 & 0 \end{pmatrix} \qquad \sigma_2 = \begin{pmatrix} 0 & -i \\ i & 0 \end{pmatrix} \qquad \sigma_3 = \begin{pmatrix} 1 & 0 \\ 0 & -1 \end{pmatrix} \tag{A-26}$$

also

$$0 = \begin{pmatrix} 0 & 0 \\ 0 & 0 \end{pmatrix} \qquad 1 = \begin{pmatrix} 1 & 0 \\ 0 & 1 \end{pmatrix} \tag{A-27}$$

The wave equation becomes

$$-\frac{\hbar}{i}\frac{\partial \psi}{\partial t} = H\psi = +\frac{\hbar c}{i}\boldsymbol{\alpha} \cdot \nabla\psi + \beta mc^2\psi \tag{A-28}$$

Since $\boldsymbol{\alpha}$ and β are 4 × 4 matrices, ψ must have four components for this equation to make sense. Equation A-28 is called the Dirac equation. It may be put in a more symmetrical form by multiplying through by β and using $\beta^2 = 1$. We then obtain

$$imc\psi = \gamma_\mu p_\mu \psi = \frac{\hbar}{i}\gamma_\mu \frac{\partial}{\partial x_\mu}\psi \tag{A-29}$$

where we use the x_4 notation and $\gamma_i = -i\beta\alpha_i$ for $i = 1, 2, 3$, $\gamma_4 = \beta$.

Next we investigate the relativistic invariance of Eq. A-29. If Eq. A-29 is to be relativistically invariant then in the new coordinate system the equation

$$(\gamma'_\mu p'_\mu - imc)\psi'(x'_\mu) = 0 \tag{A-30}$$

must be true. Here the prime denotes the variables in the transformed coordinate system. We assume that since the elements of the matrices are pure number they remain unchanged. Let

$$\psi'(x'_\mu) = S\psi(x_\mu) \tag{A-31}$$

where S is some matrix still to be determined. It follows from Eq. A-19 that

$$p'_\mu = a_{\mu\lambda}p_\lambda \tag{A-32}$$

so that Eq. A-30 can be written as

$$(\gamma_\mu a_{\mu\lambda}p_\lambda - imc)S\psi - 0 \tag{A-33}$$

Multiplying from the left with S^{-1} gives

$$(S^{-1}\gamma_\mu S a_{\mu\lambda} p_\lambda - imc)\psi = 0 \tag{A-34}$$

Comparing this with Eq. A-29 shows that the two equations are the same if

$$S^{-1}\gamma_\mu S a_{\mu\lambda} = \gamma_\lambda \tag{A-35}$$

which can also be written as

$$S^{-1}\gamma_\mu S = a_{\mu\lambda}\gamma_\lambda \tag{A-36}$$

This gives a nonlinear equation for the determination of the elements of S.

The Dirac wavefunctions do not transform as tensors of any rank; instead they are what are called spinors. Certain bilinear combinations of ψ's do transform as tensors. Let us write

$$\psi = \begin{bmatrix} \psi_1 \\ \psi_2 \\ \psi_3 \\ \psi_4 \end{bmatrix}, \qquad \psi^+ = [\psi_1^*, \psi_2^*, \psi_3^*, \psi_4^*] \tag{A-37}$$

and

$$\bar{\psi} = i\psi^+\gamma_4 \tag{A-38}$$

and similarly for some other spinor ϕ. Then it can be shown that

$$\bar{\psi}\phi \tag{A-39}$$

is a scalar,

$$\bar{\psi}\gamma_5\phi \tag{A-40}$$

is a pseudoscalar ($\gamma_5 \equiv \gamma_1\gamma_2\gamma_3\gamma_4$),

$$\bar{\psi}\gamma_\mu\phi$$

is a 4-vector,

$$\bar{\psi}\gamma_2\gamma_3\gamma_4\phi, \; \bar{\psi}\gamma_3\gamma_1\gamma_4\phi, \; \bar{\psi}\gamma_1\gamma_2\gamma_4\phi \tag{A-41}$$

and $\bar{\psi}\gamma_1\gamma_2\gamma_3\phi$ are the components of an axial 4-vector, and

$$\begin{gathered} \bar{\psi}\gamma_2\gamma_3\phi, \; \bar{\psi}\gamma_3\gamma_1\phi, \; \bar{\psi}\gamma_1\gamma_2\phi \\ \bar{\psi}\gamma_1\gamma_4\phi, \; \bar{\psi}\gamma_2\gamma_4\phi, \; \bar{\psi}\gamma_3\gamma_4\phi \end{gathered} \tag{A-42}$$

are the six components of a antisymmetric tensor.

Problem A-2. Consider the infinitesimal Lorentz transformation

$$a_{\mu\nu} = \delta_{\mu\nu} + \varepsilon_{\mu\nu}$$

where $\varepsilon_{\mu\nu}$ is infinitesimal. Show from $a_{\mu\nu}a_{\mu\lambda} = \delta_{\mu\nu}$ that $\varepsilon_{\lambda\nu} = -\varepsilon_{\nu\lambda}$. Write $S = 1 + T$ where T is of order $\varepsilon_{\mu\nu}$. Use Eq. A-36 to show that

$$S = 1 + \tfrac{1}{4}\varepsilon_{\mu\nu}\gamma_\mu\gamma_\nu = 1 + T \tag{A-43}$$

This may be used to find S for infinitesimal transformations and then by iteration S can be found for finite transformations. For example consider the infinitesimal rotation about the z-axis:

$$x_4' = x_4, \qquad x_3' = x_3, \qquad x_1' = x_1 + \varepsilon x_2, \qquad x_2' = x_2 - \varepsilon x_1 \qquad \text{(A-44)}$$

Show that

$$S_\varepsilon = 1 + \frac{\varepsilon}{2}\gamma_1\gamma_2 = \begin{bmatrix} 1 + \frac{i}{2}\varepsilon & 0 & 0 & 0 \\ 0 & 1 - \frac{i}{2}\varepsilon & 0 & 0 \\ 0 & 0 & 1 + \frac{i}{2}\varepsilon & 0 \\ 0 & 0 & 0 & 1 - \frac{i}{2}\varepsilon \end{bmatrix} \qquad \text{(A-45)}$$

By iteration show that for a finite rotation through an angle ϕ

$$S_\phi = \begin{bmatrix} e^{i(\phi/2)} & 0 & 0 & 0 \\ 0 & e^{-i(\phi/2)} & 0 & 0 \\ 0 & 0 & e^{i(\phi/2)} & 0 \\ 0 & 0 & 0 & e^{-i(\phi/2)} \end{bmatrix} \qquad \text{(A-46)}$$

Note that for $\phi = 2\pi$, $\psi' = -\psi$. This would be unsatisfactory if ψ itself were an observable. However, ψ always enters quadratically into any observable quantity.

As another example consider the infinitesimal Lorentz transformation:

$$\begin{aligned} x_1' &= x_1 - \varepsilon tc = x_1 + i\varepsilon x_4 \\ x_4' &= x_4 - i\varepsilon x_1 \\ x_2' &= x_2 \\ x_2' &= x_2 \end{aligned} \qquad \text{(A-47)}$$

Show that

$$S_\varepsilon = 1 - \frac{i\varepsilon}{2}\gamma_1\gamma_4 = 1 + \frac{\varepsilon}{2}\alpha_1$$

and that for a finite Lorentz transformation

$$S_\beta = \sqrt{\frac{1 + \sqrt{1 - \beta^2}}{2\sqrt{1 - \beta^2}}} + \alpha_1\sqrt{\frac{1 - \sqrt{1 - \beta^2}}{2\sqrt{1 - \beta^2}}} \qquad \text{(A-48)}$$

where $\beta = v/c$.

Problem A-3. Show that for the space reflection transformation $x_i' = -x_i$ $(i = 1, 2, 3)$, $x_4' = x_4$ the transformation matrix is

$$S_{\text{ref}} = \gamma_4 = \beta \tag{A-49}$$

Our next task is to solve the Dirac equation for a free particle. We take the equation in the form given in Eq. A-28. It is natural to look for a plane wave solution

$$\psi(\mathbf{x}, t) = ue^{i/\hbar(\mathbf{p}\cdot\mathbf{x}-Et)} \tag{A-50}$$

where u is a 4-component spinor. Equation A-28 becomes

$$(H - E)u = [c\alpha \cdot \mathbf{p} + \beta mc^2 - E]u = 0 \tag{A-51}$$

Equation A-51 is four linear homogeneous equations for u. The condition that a nontrivial solution exist is that the determinant of the coefficients vanish. It is easily shown that this gives

$$E = \pm\sqrt{c^2p^2 + m^2c^4} \tag{A-52}$$

With these values for E the set of equations can be solved for the components of u. Four column vectors are obtained. Two correspond to the positive sign of E, and two correspond to the negative sign of E. The solutions for u is a little complicated because of this degeneracy. A simple shortcut for finding the solutions is the following. We note that

$$(H - E)(H + E) = H^2 - E^2 = 0 \tag{A-53}$$

since $H^2 = c^2p^2 + m^2c^4$ by Eq. A-23 and E is given by Eq. A-52. Now $H + E$ is the matrix

$$(H + E) = \begin{bmatrix} E + mc^2 & 0 & cp_z & cp_- \\ 0 & E + mc^2 & cp_+ & -cp_z \\ cp_z & cp_- & E - mc^2 & 0 \\ cp_+ & -cp_z & 0 & E - mc^2 \end{bmatrix} \tag{A-54}$$

where $p_\pm = p_x \pm ip_y$.

From Eq. A-53 we see that each column of $H + E$ will give zero when operated on by (H $-$ E). Therefore, the columns of $H + E$ are the solutions we are looking for. We then multiply each column by a factor which normalizes it properly. In this way we find the four solutions

$$u^{(1)} = \sqrt{\frac{R + mc^2}{2R}} \begin{bmatrix} 1 \\ 0 \\ cp_z/(R + mc^2) \\ cp_+/(R + mc^2) \end{bmatrix}$$

$$u^{(2)} = \sqrt{\frac{R + mc^2}{2R}} \begin{bmatrix} 0 \\ 1 \\ cp_-/(R + mc^2) \\ -cp_z/(R + mc^2) \end{bmatrix}$$

$$u^{(3)} = \sqrt{\frac{R + mc^2}{2R}} \begin{bmatrix} -cp_z/(R + mc^2) \\ -cp_+/(R + mc^2) \\ 1 \\ 0 \end{bmatrix} \tag{A-55}$$

$$u^{(4)} = \sqrt{\frac{R + mc^2}{2R}} \begin{bmatrix} -cp_-/(R + mc^2) \\ cp_z/(R + mc^2) \\ 0 \\ 1 \end{bmatrix}$$

We have let $R = |E| = +\sqrt{c^2p^2 + m^2c^4}$. The normalization is chosen so that

$$u^{(\mu)^+} u^{(\nu)} = \delta_{\mu\nu} \tag{A-56}$$

The solutions $u^{(1)}$ and $u^{(2)}$ correspond to $E = +R$, and the solutions $u^{(3)}$ and $u^{(4)}$ correspond to $E = -R$.

It may be shown (see Problem A-4) that the Dirac equation describes particles of spin $-\frac{1}{2}$. Solutions $u^{(1)}$ and $u^{(3)}$ correspond to the orientation of this spin along the $+z$-axis while $u^{(2)}$ and $u^{(4)}$ correspond to its orientation along the $-z$-axis.

Problem A-4. Show that the orbital angular momentum operator

$$\mathbf{L} = \mathbf{x} \times \mathbf{p} \tag{A-57}$$

does not commute with the Dirac Hamiltonian $H = c\boldsymbol{\alpha} \cdot \mathbf{p} + \beta mc^2$, but that

$$\mathbf{J} = \mathbf{L} + \mathbf{S} \tag{A-58}$$

where

$$\mathbf{S} = \frac{\hbar}{2i}(\alpha_2\alpha_3, \alpha_3\alpha_1, \alpha_1\alpha_2) \tag{A-59}$$

does commute with H. $\mathbf{S}$ may be interpreted as the spin operator. Show that the eigenvalues of any component of S are $\pm\hbar/2$. Show that the eigenvalues of S^2 are $3\hbar^2/4$.

Problem A-5. Write the 4-component Dirac spinor as

$$\Psi = \begin{bmatrix} \phi \\ \chi \end{bmatrix} \tag{A-60}$$

where ϕ and χ are 2-component spinors. Show that ϕ and χ obey the coupled equations

$$\begin{aligned} -\frac{\hbar}{i}\frac{\partial\phi}{\partial t} &= c\left(\frac{\hbar}{i}\nabla - \frac{e}{c}\mathbf{A}\right) \cdot \boldsymbol{\sigma}\chi + (e\Phi + mc^2)\phi \\ -\frac{\hbar}{i}\frac{\partial\chi}{\partial t} &= c\left(\frac{\hbar}{i}\nabla - \frac{e}{c}\mathbf{A}\right) \cdot \boldsymbol{\sigma}\phi + (e\Phi - mc^2)\chi \end{aligned} \tag{A-61}$$

in the presence of an electromagnetic field described by the potentials $\mathbf{A}$ and Φ. Show that in the nonrelativistic limit χ can be eliminated and ϕ satisfies the equation

$$-\frac{\hbar}{i}\frac{\partial\phi}{\partial t} = \frac{1}{2m}\left(\frac{\hbar}{i}\nabla - \frac{e}{c}\mathbf{A}\right)^2\phi + (e\Phi + mc^2)\phi - \frac{e\hbar}{2mc}\mathbf{B} \cdot \boldsymbol{\sigma}\phi \tag{A-62}$$

This shows that the Dirac electron has a magnetic moment of

$$\boldsymbol{\mu} = -\frac{e\hbar}{2mc}\boldsymbol{\sigma} \tag{A-63}$$

Problem A-6. A 2-component theory of the neutrino has been proposed by Yang and Lee. The Hamiltonian is taken to be

$$H = -c\boldsymbol{\sigma} \cdot \mathbf{p} \tag{A-64}$$

(a) Find the eigenvalues and eigenfunctions of H.
(b) Show that $\mathbf{L}$ does not commute with H but that

$$\mathbf{J} = \mathbf{L} + \frac{\hbar}{2}\boldsymbol{\sigma} \tag{A-65}$$

does commute.

(c) Show that a positive energy neutrino has its spin antiparallel to its momentum.

Appendix B

Details of the Calculation of the Klein-Nishina Cross Section

Our purpose here is to calculate M_{fi} of Eq. 6-44 and use it in the calculation of the cross section for Compton scattering. The necessary matrix elements are given in Eq. 6-46. We can simplify the notation somewhat by denoting the Dirac spinors $u_{\mathbf{q}_i\lambda_f}$ and $u_{\mathbf{q}_f\lambda_i}$ by u_i and u_f. Also we shall denote $\boldsymbol{\alpha} \cdot \mathbf{u}_i$ and $\boldsymbol{\alpha} \cdot \mathbf{u}_f$ by α_i and α_f. Then Eq. 6-44 may be written as

$$M_{fi} = \frac{e^2}{\sqrt{\omega_i \cdot \omega_f}} \left(\frac{2\pi\hbar c^2}{\Omega}\right) \sum_\lambda \left\{ \frac{(u_f^+\alpha_f u_1)(u_1^+\alpha_i u_i)}{E_i - E_{I1}} + \frac{(u_f^+\alpha_i u_2)(u_2^+\alpha_f u_i)}{E_i - E_{I2}} \right\} \tag{B-1}$$

The sum over λ is a sum over the spins and signs of the energies of the intermediate states. We write

$$M_{fi} = \frac{e^2}{\sqrt{\omega_i \cdot \omega_f}} \left(\frac{2\pi\hbar c^2}{\Omega}\right) M'_{fi} \tag{B-2}$$

In evaluating M'_{fi} it is convenient to choose units so that $\hbar = c = 1$. The original units are easily restored by noting that M'_{fi} has units of (energy)$^{-1}$. In these units

$$E_i = m + k_i \tag{B-3a}$$

$$E_{I1} = \pm\sqrt{k_i^2 + m^2} = E_1 \tag{B-3b}$$

$$E_{I2} = \pm\sqrt{k_f^2 + m^2} + k_i + k_f = E_2 + k_i + k_f \tag{B-3c}$$

We have let $q_i = 0$, so the electron is initially at rest. We have denoted the energies of the electron in the intermediate states by E_1 and E_2. The M'_{fi} is given by

$$M'_{fi} = \sum_\lambda \left\{ \frac{(u_f^+\alpha_f u_1)(u_1^+\alpha_i u_i)}{m + k_i - E_1} + \frac{(u_f^+\alpha_i u_2)(u_2^+\alpha_f u_i)}{m - k_f - E_2} \right\} \tag{B-4}$$

We cannot remove the denominators from the sum because both signs of E_1 and E_2 occur in the sum over λ. However, if we multiply numerator and denominator of the first term by $m + k_i + E_1$ and the numerator and denominator of the second term by $m - k_f + E_2$ then the denominators can be extracted and we obtain

$$M'_{fi} = \frac{1}{(m + k_i)^2 - E_1^2} \sum_\lambda (m + k_i + E_1)(u_f \alpha_f u_1)(u_1 \alpha_i u_i)$$
$$+ \frac{1}{(m - k_f)^2 - E_2^2} \sum_\lambda (m - k_f + E_2)(u_f \alpha_i \cdot u_2)(u_2 \alpha_f u_i) \quad \text{(B-5)}$$

This can be simplified by noting that

$$(m + k_i)^2 - E_1 = 2mk_i \quad \text{(B-6a)}$$
$$(m - k_f)^2 - E_2^2 = -2mk_f \quad \text{(B-6b)}$$

We use

$$E_1 u_1 = H_1 u_1 = (\mathbf{k}_i \cdot \boldsymbol{\alpha} + m\beta) u_1 \quad \text{(B-7a)}$$

and

$$E_2 u_2 = H_2 u_2 = (-\mathbf{k}_f \cdot \boldsymbol{\alpha} + m\beta) u_2 \quad \text{(B-7b)}$$

Just as in the section on Čerenkov radiation, we can use the completeness relation to obtain

$$\sum_{\lambda=1}^{4} u_1 u_1^{+} = \sum_{\lambda=1}^{4} u_2 u_2^{+} = 1 \quad \text{(B-8)}$$

where 1 is the 4 × 4 matrix. In this way we obtain

$$M'_{fi} = \frac{1}{2m} (u_f^{+} Q u_i) \quad \text{(B-9a)}$$

where

$$Q = \frac{1}{k_i} \alpha_f (m + k_i + \boldsymbol{\alpha} \cdot \mathbf{k}_i + \beta m) \alpha_i$$
$$- \frac{1}{k_f} \alpha_i (m - k_f - \boldsymbol{\alpha} \cdot \mathbf{k}_f + \beta m) \alpha_f \quad \text{(B-9b)}$$

Now, $\beta \alpha_i = -\alpha_i \beta$ and $\beta \alpha_f = -\alpha_f \beta$, so that the β can be moved to the right in Q. Then

$$\beta u_i = u_i \quad \text{(B-10)}$$

since $\mathbf{q}_i = 0$, and the terms containing βm cancel the terms containing m in Q. Next, we use Eq. 6-30 to obtain

$$\alpha_f \alpha_i + \alpha_i \alpha_f = 2(\mathbf{u}_i \cdot \mathbf{u}_f) 1 \quad \text{(B-11)}$$

and obtain

$$Q = 2\mathbf{u}_i \cdot \mathbf{u}_f + \frac{1}{k_i} \alpha_f (\boldsymbol{\alpha} \cdot \mathbf{k}_i) \alpha_i + \frac{1}{h_f} \alpha_i (\boldsymbol{\alpha} \cdot \mathbf{k}_f) \alpha_f \quad \text{(B-12)}$$

The differential cross section is proportional to the square of M'_{fi}. We are not interested in the spin of the electron in the initial or final states; so we sum $|M'_{fi}|^2$ over final spin states and average over initial spin states. The quantity we want is

$$\frac{1}{2}\sum_{\lambda_i=1}^{2}\sum_{\lambda_f=1}^{2}\frac{1}{4m^2}|u_f^+Qu_i|^2 = \frac{1}{2}\sum_{\lambda_i=1}^{2}\sum_{\lambda_f=1}^{2}\frac{1}{4m^2}(u_f^+Qu_i)(u_i^+Q^+u_f) \qquad \text{(B-13)}$$

We can extend the sums over λ_i and λ_f by using

$$\frac{H_i + |E_i|}{2\,|E_i|}\,u_i = \begin{cases} u_i & \lambda_i = 1, 2 \\ 0 & \lambda_i = 3, 4 \end{cases} \qquad \text{(B-14a)}$$

$$\frac{H_f + |E_f|}{2\,|E_f|}\,u_f = \begin{cases} u_f & \lambda_f = 1, 2 \\ 0 & \lambda_f = 3, 4 \end{cases} \qquad \text{(B-14b)}$$

$$H_i = \beta m, \qquad E_i = m \qquad \text{(B-14c)}$$

$$H_f = (\mathbf{k}_i - \mathbf{k}_f)\cdot\boldsymbol{\alpha} + \beta m, \qquad E_f = m + k_i - k_f \qquad \text{(B-14d)}$$

Equation B-13 becomes

$$\frac{1}{32m^2\,|E_i|\,|E_f|}\sum_{\lambda_i=1}^{4}\sum_{\lambda_f=1}^{4}(u_f^+Q(H_i + |E_i|)u_i)(u_i^+Q^+(H_f + |E_f|)u_f)$$

$$= \frac{1}{32m^2\,|E_i|\,|E_f|}\operatorname{Tr} Q(H_i + |E_i|)Q^+(H_f + |E_f|) \qquad \text{(B-15)}$$

From this point on some tedious algebra is unavoidable. It can be reduced to a minimum by simplifying the notation and using the properties of the Dirac matrices. Let

$$\mathbf{u}_1 = \frac{\mathbf{k}_i}{k_i}, \qquad \mathbf{u}_2 = \frac{\mathbf{k}_f}{k_f}$$

$$\alpha_1 = \mathbf{u}_1\cdot\boldsymbol{\alpha}, \qquad \alpha_2 = \mathbf{u}_2\cdot\boldsymbol{\alpha} \qquad \text{(B-16)}$$

Equation B-12 becomes

$$Q = 2(\mathbf{u}_i\cdot\mathbf{u}_f) + \alpha_f\alpha_1\alpha_i + \alpha_i\alpha_2\alpha_f \qquad \text{(B-17)}$$

Equation B-15 becomes

$$\text{(B-15)} = \frac{1}{32m^2\,|E_i|\,|E_f|}\operatorname{Tr}(2\mathbf{u}_i\cdot\mathbf{u}_f + \alpha_f\alpha_1\alpha_i + \alpha_i\alpha_2\alpha_f)$$

$$\times (m\beta + m)(2\mathbf{u}_i\cdot\mathbf{u}_f + \alpha_i\alpha_1\alpha_f + \alpha_f\alpha_2\alpha_1)$$

$$\times (k_i\alpha_1 - k_f\alpha_2 + \beta m + k_i - k_f + m) \qquad \text{(B-18)}$$

Now it is easily proven that

$$\text{Tr}\ \alpha_i\alpha_j \cdots \alpha_n = 0 \tag{B-19}$$

when there are an odd number of the α's in the product. Similarly, the trace of such a product vanishes when there are an odd number of β's among the factors. When there is an even number of β's, one may use $\alpha_i\beta = -\beta\alpha_i$ to move the β's together and then use $\beta^2 = 1$. In this way many of the terms in Eq. B-18 may be shown to be zero. What remains may be reduced by using

$$\alpha_1\alpha_i + \alpha_i\alpha_1 = 2(\mathbf{u}_1 \cdot \mathbf{u}_i)1 = 0 \tag{B-20a}$$

$$\alpha_2\alpha_f + \alpha_f\alpha_2 = 2(\mathbf{u}_2 \cdot \mathbf{u}_f)1 = 0 \tag{B-20b}$$

$$\alpha_1{}^2 = \alpha_2{}^2 = \alpha_i{}^2 = \alpha_f{}^2 = 1 \tag{B-20c}$$

Equation B-18 reduces to

$$\frac{1}{32m^2\,|E_i|\,|E_f|}\ \text{Tr}\ \{8m^2(\mathbf{u}_i \cdot \mathbf{u}_f)^2 + m(k_i - k_f) \times [-2(\mathbf{u}_i \cdot \mathbf{u}_f)(\alpha_1\alpha_2\alpha_f\alpha_i + \alpha_2\alpha_1\alpha_i\alpha_f) + 2 + \alpha_f\alpha_1\alpha_i\alpha_f\alpha_2\alpha_i + \alpha_i\alpha_2\alpha_f\alpha_i\alpha_1\alpha_f]\} \tag{B-21}$$

We now use

$$\text{Tr}\ (\alpha_1\alpha_2\alpha_f\alpha_i + \alpha_2\alpha_1\alpha_i\alpha_f) = 8(\mathbf{u}_i \cdot \mathbf{u}_f)(\mathbf{u}_1 \cdot \mathbf{u}_2) - 8(\mathbf{u}_i \cdot \mathbf{u}_2)(\mathbf{u}_1 \cdot \mathbf{u}_f) \tag{B-22a}$$

$$\text{Tr}\ (\alpha_f\alpha_i\alpha_f\alpha_i\alpha_1\alpha_2 + \alpha_i\alpha_f\alpha_i\alpha_f\alpha_2\alpha_1) = 16(\mathbf{u}_i \cdot \mathbf{u}_f)^2(\mathbf{u}_1 \cdot \mathbf{u}_2) - 16(\mathbf{u}_i \cdot \mathbf{u}_f)(\mathbf{u}_1 \cdot \mathbf{u}_f)(\mathbf{u}_2 \cdot \mathbf{u}_i) - 2(\mathbf{u}_1 \cdot \mathbf{u}_2) \tag{B-22b}$$

and

$$\mathbf{u}_1 \cdot \mathbf{u}_2 = \cos\theta = 1 - \frac{mc}{\hbar}\left(\frac{1}{k_f} - \frac{1}{k_i}\right) \tag{B-23c}$$

to obtain

$$(\text{B-13}) = \frac{1}{|E_i|\,|E_f|}\left\{(\mathbf{u}_i \cdot \mathbf{u}_f) - \frac{1}{2} + \frac{1}{4}\frac{k_i}{k_f} + \frac{1}{4}\frac{k_f}{k_i}\right\} \tag{B-24}$$

When this is used in the Fermi golden rule to calculate the cross section, The Klein-Nishina formula, Eq. 6-53, is obtained.

Appendix C

Answers and Solutions to the Problems

CHAPTER 1

Problem 1-1. Write

$$\begin{aligned}
\operatorname{Tr} C &= \sum_{A'} \langle A'|\, C\, |A'\rangle = \operatorname{Tr} 1C1 \\
&= \sum_{A'} \sum_{B'} \sum_{B''} \langle A' \mid B'\rangle \langle B'|\, C\, |B''\rangle \langle B'' \mid A'\rangle \\
&= \sum_{A'} \sum_{B'} \sum_{B''} \langle B'' \mid A'\rangle \langle A' \mid B'\rangle \langle B'|\, C\, |B''\rangle \\
&= \sum_{B'} \sum_{B''} \langle B'' \mid B'\rangle \langle B'|\, C\, |B''\rangle \\
&= \sum_{B'} \langle B'|\, C\, |B''\rangle
\end{aligned}$$

where Eqs. 1-41 and 1-32 have been used.

Problem 1-2. Write

$$\begin{aligned}
\sum_{A'} \sum_{A''} |\langle A'|\, C\, |A''\rangle|^2 &= \sum_{A'} \sum_{A''} \langle A'|\, C\, |A''\rangle \langle A'|\, C\, |A''\rangle^* \\
&= \sum_{A'} \sum_{A''} \langle A'|\, C\, |A''\rangle \langle A''|\, C^+\, |A'\rangle \\
&= \sum_{A'} \langle A'|\, CC^+\, |A'\rangle = \operatorname{Tr} CC^+
\end{aligned}$$

where Eqs. 1-27 and 1-43 have been used.

Problem 1-3. Write

$$\begin{aligned}
\langle B'|\, f(A)\, |B'\rangle &= \sum_{A',} \sum_{A''} \langle B' \mid A'\rangle \langle A'|\, f(A)\, |A''\rangle \langle A'' \mid B'\rangle \\
&= \sum_{A'} \sum_{A''} \langle B' \mid A'\rangle f(A') \delta_{A',A''} \langle A'' \mid B'\rangle \\
&= \sum_{A'} \langle B' \mid A'\rangle f(A') \langle A' \mid B'\rangle
\end{aligned}$$

where Eqs. 1-41 and $f(A)\,|A'\rangle = f(A')\,|A'\rangle$ have been used.

Problem 1-4. Use the power series for the exponential to write

$$e^{(i\beta\sigma_x/2)} = \sum_{n=0}^{\infty} \frac{1}{n!}\left(\frac{i\beta}{2}\right)^n \sigma_x{}^n$$

Note that $\sigma_x{}^2 = 1$. It follows that

$$e^{(i\beta\sigma_x/2)} = 1 \sum_{n \text{ even}} \frac{1}{n!}\left(\frac{i\beta}{2}\right)^n + \sigma_x \sum_{n \text{ even}} \frac{1}{n!}(i\beta)^n$$

Recognizing the series for cos $\beta/2$ and sin $\beta/2$ we see that

$$e^{(i\beta\sigma_x/2)} = 1 \cos\frac{\beta}{2} + i\sigma_x \sin\frac{\beta}{2}$$

which is the same as Eq. 1-53. This problem can also be solved by using Eq. 1-51. It is convenient to define the vectors

$$|z, +1\rangle = \begin{pmatrix}1\\0\end{pmatrix} \qquad \text{and} \qquad |z, -1\rangle = \begin{pmatrix}0\\1\end{pmatrix}$$

(Actually, these are the eigenvectors of σ_z given by Eq. 1-85.) Then

$$\langle z, i|\ \sigma_x\ |z, j\rangle = \begin{pmatrix}0 & 1\\1 & 0\end{pmatrix}$$

Solving the eigenvalue problem

$$\sigma_x\ |x, \lambda\rangle = \lambda\ |x, \lambda\rangle$$

we find that $\lambda = \pm 1$ and the normalized eigenvectors are

$$|x, +1\rangle = \frac{1}{\sqrt{2}}\begin{pmatrix}1\\1\end{pmatrix}, \qquad |x, -1\rangle = \frac{1}{\sqrt{2}}\begin{pmatrix}1\\-1\end{pmatrix}$$

Then Eq. 1-51 gives

$$\langle z, i|\ e^{(i\beta\sigma_x/2)}\ |z, j\rangle = \sum_{\lambda=\pm 1} \langle z, i \mid x, \lambda\rangle e^{(i\beta\lambda/2)} \langle x, \lambda \mid z, j\rangle$$

This gives Eq. 1-53. For instance setting $i = j = 1$ gives

$$\tfrac{1}{2}e^{i\beta/2} + \tfrac{1}{2}e^{-i\beta/2} = \cos\frac{\beta}{2}$$

which is the $i = 1, j = 1$ element of the matrix in Eq. 1-53.

Problem 1-5. Writing $L_x = yp_z - zp_y$ and $L_y = zp_x - xp_z$ we find

$$\begin{aligned}[L_x, L_y] &= (yp_z - zp_y)(zp_x - xp_z) - (zp_x - xp_z)(yp_z - zp_y)\\ &= yp_x(p_z z - zp_z) + xp_y(zp_z - p_z z)\\ &= ih(xp_y - yp_x) = i\hbar L_z\end{aligned}$$

Problem 1-6. We get the vector

$$|\uparrow, x\rangle = \frac{1}{\sqrt{2}}\begin{pmatrix}1\\1\end{pmatrix}$$

from Eq. 1-91a by setting $\theta = \pi/2$ and $\phi = 0$. Then

$$\langle\uparrow, x \mid \psi_t\rangle = \tfrac{1}{2}(1, 1)\begin{pmatrix}e^{-i\omega t}\\e^{+i\omega t}\end{pmatrix}$$
$$= \cos \omega t$$

The other relations are obtained in the same way.

Problem 1-7. Write

$$\langle p'|\, xp - px\,|p''\rangle = \int d'''p\{\langle p'|\, x\,|p'''\rangle\langle p'''|\, p\,|p''\rangle - \langle p'|\, p\,|p'''\rangle\langle p'''|\, x\,|p''\rangle\}$$
$$= i\hbar\, \delta(p' - p'')$$
$$= (p'' - p')\langle p'|\, x\,|p''\rangle$$

Equation 1-109 follows from this just as Eq. 1-108 followed from Eq. 1-107.

Problem 1-8. Let

$$|c\rangle = a^+\,|a\rangle$$

Then

$$n\,|c\rangle = a^+aa^+\,|a\rangle = a^+(a^+a + 1)\,|n\rangle = (n + 1)\,|c\rangle$$

It follows that

$$|c\rangle = D_n\,|n + 1\rangle$$

The normalization constant is found to be $\sqrt{n + 1}$ times an arbitrary phase factor.

Problem 1-9. Use Eq. 1-154a to write

$$\langle n_1|\, x^2\,|n_2\rangle$$
$$= \sum \langle n_1|\, x\,|n\rangle\langle n|\, x\,|n_2\rangle$$
$$= \frac{\hbar}{2m\omega}\sum_n \{\sqrt{n + 1}\,\delta_{n_1, n+1} + \sqrt{n_2}\,\delta_{n_1, n-1}\}\{\sqrt{n + 1}\,\delta_{n, n_2+1} + \sqrt{n_2}\,\delta_{n, n_2-1}\}$$
$$= \frac{\hbar}{2m\omega}\{(2n_1 + 1)\delta_{n_1, n_2} + \sqrt{n_1}\sqrt{n_2 + 1}\,\delta_{n_1, n_2+2} + \sqrt{n_1 + 1}\sqrt{n_2}\,\delta_{n_1, n_2-2}\}$$

A similar calculation using Eq. 1-154b gives

$$\langle n_1|\, p^2\,|n_2\rangle$$
$$= \frac{m\hbar\omega}{2}\{(2n_1 + 1)\delta_{n_1 n_2} - n_1\sqrt{n_2 + 1}\,\delta_{n_1, n_2+2} - \sqrt{n_1 + 1}\sqrt{n_2}\,\delta_{n_1, n_2-2}\}$$

It follows that

$$\frac{1}{2m}\langle n_1|\, p^2\, |n_2\rangle + \frac{m\omega^2}{2}\langle n_1|\, x^2\, |n_2\rangle = \hbar\omega(n + \tfrac{1}{2})$$

CHAPTER 2

Problem 2-1. Write

$$\begin{aligned}
\langle c|\, a^+a\, |c\rangle &= \sum_{m=0}^{\infty}\sum_{n=0}^{\infty} b_m^* b_n \langle m|\, a^+a\, |n\rangle \\
&= \sum_{m=0}^{\infty}\sum_{n=0}^{\infty} b_m^* b_n n \delta_{n,m} \\
&= \sum_{n=0}^{\infty} \frac{|c|^n\, e^{-|c|^2}}{n!}\, n \\
&= |c|^2\, e^{-|c|^2} \sum_{n=0}^{\infty} \frac{|c|^{n-1}}{(n-1)!} \\
&= |c|^2
\end{aligned}$$

Equations 2-39b to e can be derived in a similar manner.

Problem 2-2. Using Eq. 2-29 we find

$$\langle c|\, E^2\, |c\rangle = \frac{2\pi\hbar\omega}{\Omega}\{\langle c|\, a^+a + aa^+\, |c\rangle - \langle c|\, a^2\, |c\rangle e^{+i2\mathbf{k}\cdot\mathbf{x}} - \langle c|\, a^{+2}\, |c\rangle e^{-i2\mathbf{k}\cdot\mathbf{x}}\}$$

The necessary expectation values are given in Eq. 2-39. Equation 2-38 gives $|\langle c|\, \mathbf{E}\, |c\rangle|^2$. The difference gives Eq. 2-42.

CHAPTER 3

Problem 3-1. We can construct a three-dimensional space with coordinate axes n_x, n_y, and n_z. Since there is a mode of the electromagnetic field with a given polarization for each triplet of integers (n_x, n_y, n_z), there must be $\Delta n_x\, \Delta n_y\, \Delta n_z$ modes with n_x in the range Δn_x, n_y in the range Δn_y, and n_z in the range Δn_z. Since $\Delta n_j = L\,\Delta k_i/2\pi$ we can say that

$$\frac{L^3}{(2\pi)^3}\Delta k_x\, \Delta k_y\, \Delta k_z$$

is the number of modes with k_x in Δk_x, and so on. Taking the limit as $L \to \infty$ and $\Delta k_i \to 0$ gives

$$\frac{\Omega}{(2\pi)^3}\, d^3k$$

for the number of modes in an infinitely large box with **k** in d^3k.

Problem 3-2. The atomic wave functions are of the form

$$\psi_{nlm}(\mathbf{x}) = R_{nl}(r)Y_l^m(\theta, \phi)$$

where $Y_l^m(\theta, \phi)$ is a spherical harmonic. We can write

$$\begin{aligned} x &= r \sin\theta \cos\phi = r(a_x Y_1^1 + b_x Y_1^{-1}) \\ y &= r \sin\theta \cos\phi = r(a_y Y_1^1 + b_y Y_1^{-1}) \\ z &= r \cos\theta \qquad\;\;\; = r a_z Y_1^0 \end{aligned}$$

where a_x, b_x, a_y, and so on, are constants. Now

$$Y_1^{\pm 1} Y_l^m = A Y_{l+1}^{m\pm 1} + B Y_{l-1}^{m\pm 1} \qquad \text{and} \qquad Y_1^0 Y_l^m = C Y_{l+1}^m + D Y_{l-1}^m$$

where A, B, C, and D are constants. We see immediately that the matrix elements of x and y vanish unless $\Delta l = \pm 1$ and $\Delta m = \pm 1$, and the matrix elements of z vanish unless $\Delta l = \pm 1$ and $\Delta m = 0$.

Problem 3-3. The matrix element for the transition from the $2p$ state with $m = 0$ to the $2s$ state is

$$\begin{aligned} \langle 2p, 0|\, \mathbf{x}\, |1s\rangle &= \langle 2p, 0|\, z\, |1s\rangle \mathbf{e} \\ &= \frac{1}{8\pi a^3}\int d^3x\, \frac{r}{a}\, e^{-\frac{3}{2}r/a} r\sqrt{2}\cos^2\theta \\ &= 4\sqrt{2}\, a(\tfrac{2}{3})^5 \end{aligned}$$

This may be used in Eq. 3-19 together with

$$\frac{\omega}{c} = \frac{\Delta E}{\hbar c} = \frac{3}{8}\frac{e^2}{a}$$

to obtain

$$\tau = \left(\frac{8}{3}\right)^2 \left(\frac{\hbar c}{e^2}\right)^4 \frac{a}{c} = 1.6 \times 10^{-9} \text{ sec}$$

The lifetime for the $2p$ states with $m = +1$ and -1 is the same. For these states the matrix elements of both x and y contribute.

Problem 3-4. Choose $\mathbf{k}$ to be in the direction of the z-axis. Then

$$\langle 2s|\, \mathbf{u}\cdot\mathbf{p}e^{-i\mathbf{k}\cdot\mathbf{x}}\,|1s\rangle = \frac{\hbar}{i}\int d^3x\,\psi_{2s}e^{-ikz}\mathbf{u}\cdot\frac{\partial}{\partial\mathbf{x}}\,\psi_{1s}$$
$$= -\frac{\hbar}{ia}\int d^3x\,\psi_{2s}\psi_{1s}\frac{\mathbf{u}\cdot\mathbf{x}}{r}\,e^{-ikz}$$

Now $\mathbf{u}$ is perpendicular to $\mathbf{k}$, so it lies in the x-y plane. Since $\psi_{2s}\psi_{1s}$ has spherical symmetry the integration over x and y gives zero. Note that this result does not depend of the dipole approximation. It holds in any order of the expansion of the exponential.

Problem 3-5. The interaction energy of a magnetic dipole, $\boldsymbol{\mu}$, with a magnetic field $\mathbf{B}$ is

$$H''' = -\boldsymbol{\mu}\cdot\mathbf{B}$$

Using $\mathbf{B} = \nabla \times \mathbf{A}$ and Eq. 2-11 gives

$$H''' = -\frac{ie\hbar}{2mc}\sum_{\mathbf{k}\sigma}\left(\frac{2\pi\hbar c^2}{\Omega\omega_k}\right)^{1/2}\boldsymbol{\sigma}\cdot(\mathbf{k}\times\mathbf{u}_{\mathbf{k}\sigma})[a_{\mathbf{k}\sigma}e^{i\mathbf{k}\cdot\mathbf{x}} - a^{+}_{\mathbf{k}\sigma}e^{-ik\cdot x}]$$

Problem 3-6. The initial and final states can be taken to be

$$|i\rangle = |1s\rangle_e\,|\uparrow\rangle_e\,|\uparrow\rangle_n\,|\text{no photons}\rangle_{\text{rad}}$$

$$|f\rangle = |1s\rangle_e\frac{1}{\sqrt{2}}(|\uparrow\rangle_e\,|\downarrow\rangle_n - |\downarrow\rangle_e\,|\uparrow\rangle_n\rangle\,|\cdots 1_{\mathbf{k}\sigma}\cdots\rangle_{\text{rad}}$$

The lifetime is given by

$$\frac{1}{\tau} = \sum_{\text{final states}}\frac{2\pi}{\hbar}\,|\langle f|\,H'''\,|i\rangle|^2\,\delta(\Delta E - \hbar ck)$$
$$= \frac{\Omega}{(2\pi)^3}\left(\frac{e\hbar}{2mc}\right)^2\left(\frac{2\pi\hbar c^2}{\Omega}\right)\frac{2\pi}{\hbar^2c^2}$$
$$\times\sum_{\sigma}\int\frac{d^3k}{k}\,\delta(k_0 - k)\,|\langle f|\,\boldsymbol{\sigma}\cdot(\mathbf{k}\times\mathbf{u}_{\mathbf{k}\sigma})a^{+}_{\mathbf{k}\sigma}e^{-i\mathbf{k}\cdot\mathbf{x}}\,|i\rangle|^2$$

where $\Delta E = \hbar ck_0$ is the energy difference of the levels and $k_0 = 2\pi/21\ \text{cm}^{-1}$. Making the dipole approximation $e^{i\mathbf{k}\cdot\mathbf{x}} \simeq 1$. The matrix element is found to be

$$-\frac{1}{\sqrt{2}}[(\mathbf{k}\times\mathbf{u}_{k\sigma})_x + i(\mathbf{k}\times\mathbf{u}_{\mathbf{k}\sigma})_y] = M$$

We can show that

$$\sum_{\sigma}|M|^2 = \tfrac{1}{2}\sum_{\sigma}[(\mathbf{k}\times\mathbf{u}_{\mathbf{k}\sigma})_x + (\mathbf{k}\times\mathbf{u}_{\mathbf{k}\sigma})_y{}^2] = \tfrac{1}{2}k^2(1+\cos^2\theta)$$

(To see this, rotate the pair of polarization vectors about **k** as an axis until one of them lies in the x-y plane.) Carrying out the integrations gives

$$\frac{1}{\tau} = \frac{1}{3}\frac{e^2\hbar}{m^2c^2}k_0^{\;3}$$

from which $\tau = 2 \times 10^{14}$ sec.

Problem 3-7. The lifetime is given by

$$\frac{1}{\tau} = \frac{\Omega}{(2\pi)^3}\left(\frac{e\hbar}{2mc}\right)^2\left(\frac{2\pi\hbar c^2}{\Omega}\right)\frac{2\pi}{\hbar^2c^2}$$

$$\times \sum_\sigma \int \frac{d^3k}{k}\,\delta(k_0 - k)\,|\langle 1s\downarrow|\, \boldsymbol{\sigma} \cdot (\mathbf{k} \times \mathbf{u}_{\mathbf{k}\sigma})\,|2s\uparrow\rangle|^2$$

$$= \frac{e^2\hbar}{8\pi m^2c^2}\int \frac{d^3k}{k}\,k^2(1 + \cos^2\theta)\,\delta(k_0 - k)\,|\langle 1s|\, e^{-i\mathbf{k}\cdot\mathbf{x}}\,|2s\rangle|^2$$

$$= \frac{2}{3}\frac{e^2\hbar}{m^2c^2}k_0^{\;3}\,|\langle 1s|\, e^{-i\mathbf{k}_0\cdot\mathbf{x}}\,|2s\rangle|^2$$

If we expand the exponential we find that the matrix element of the first two terms in the expansion vanishes and

$$\langle 1s|\, e^{-i\mathbf{k}_0\cdot\mathbf{x}}\,|2s\rangle \simeq \frac{-1}{2}\langle 1s|\,(\mathbf{k}_0 \cdot \mathbf{x})^2\,|2s\rangle \simeq -\frac{k_0^{\;2}}{6}\langle 1s|\, r^2\,|2s\rangle$$

$$\frac{1}{\tau} = \frac{1}{54}\frac{e^2\hbar}{m^2c^2}k_0^{\;7}\,|\langle 1s|\, r^2\,|2s\rangle|^2$$

Approximating the matrix element of r^2 by a^2, this becomes

$$\frac{1}{\tau} = \frac{1}{54}\frac{e^2\hbar}{m^2c^2}k_0^{\;3}(k_0a)^4$$

Using $e^2/\hbar c = \frac{1}{137}$ this may be written as

$$\frac{1}{\tau} = \frac{1}{54}\left(\frac{c}{a}\right)\left(\frac{3}{8}\right)^7\frac{1}{(137)^{10}}$$

from which

$$\tau = 2 \times 10^7 \text{ sec}$$

Problem 3-8. The transition probability per unit time is

$$\frac{2\pi}{\hbar}\left(\frac{e}{mc}\right)^2\left(\frac{2\pi\hbar c^2}{\Omega\omega_k}\right)n_{\mathbf{k}\sigma}\,|\langle\mathbf{q}|\,\mathbf{p} \cdot \mathbf{u}_{\mathbf{k}\sigma}e^{+i\mathbf{k}\cdot\mathbf{x}}\,|1s\rangle|^2\,\delta\left(\frac{\hbar^2q^2}{2m} - E_{1s} - \hbar\omega_k\right)$$

where

$$\psi_{\mathbf{q}} = \langle \mathbf{x} \mid \mathbf{q}\rangle = \frac{e^{i\mathbf{q}\cdot\mathbf{x}}}{\sqrt{\Omega}}$$

is the wave function of the ejected electron and

$$\psi_{1s}(\mathbf{x}) = \langle \mathbf{x} \mid 1s \rangle = \frac{1}{\sqrt{\pi a^3}} e^{-r/a}$$

is the wave function of the electron in the ground state of hydrogen. When the transition probability per unit time is summed over all final states of the ejected electron, the result is equal to the total cross section times the flux $n_{\mathbf{k}\sigma}c/\Omega$. In this way we obtain

$$\sigma = \frac{\Omega}{2\pi} \frac{e^2}{m^2\omega_k c} \int q^2\, dq\, d\Omega_e \left| \langle \mathbf{q} | \frac{\hbar}{i} \frac{\partial}{\partial z} |1s\rangle \right|^2 \delta\left[\frac{\hbar^2 q^2}{2m} - E_{1s} - \hbar\omega\right]$$

and

$$\frac{d\sigma}{d\Omega_e} = \frac{\Omega}{2\pi} r_e \frac{q}{ka^2} |\langle \mathbf{q} | \cos\theta\, |1s\rangle|$$

We can use

$$e^{i\mathbf{q}\cdot\mathbf{x}} = 4\pi \sum_{l=0}^{\infty} \sum_{m=-l}^{+l} i^l j_l(qr) Y_{lm}(\theta, \phi) Y_{lm}(\theta'\phi')$$

where θ, ϕ and θ', ϕ' are the angles of $\mathbf{q}$ and $\mathbf{x}$ respectively. Also

$$\cos\theta' = \left(\frac{3}{4\pi}\right)^{1/2} Y_{10}(\theta', \phi)$$

We find

$$\langle \mathbf{q} | \cos\theta\, |1s\rangle = \frac{4\pi i \cos\theta}{\sqrt{\pi a^3 \Omega}} \int_0^\infty r^2\, dr j_1(qr) e^{-r/a} = 4\pi i \sqrt{\frac{a^3}{\pi\Omega}} \cos\theta f(qa)$$

where

$$f(qa) = \int_0^\infty x^2\, dx j_1(qax) e^{-x} = \frac{2qa}{(1 + q^2a^2)^2}$$

This gives

$$\frac{d\sigma}{d\Omega} = 8a^2 \left(\frac{e^2}{\hbar c}\right)^2 \cos^2\theta \frac{q}{k} f^2(qa)$$

where $\hbar q$ is the momentum and θ is the direction of the ejected electron.

Problem 3-9. Assuming $q_i = 0$, the conservation laws become

$$\hbar c k_i = \hbar c k_f + \frac{\hbar^2 q^2 f}{2m}$$

$$\mathbf{k}_i = \mathbf{k}_f + \mathbf{q}_f$$

Eliminate q_f and solve for $\lambda_f - \lambda_i$ with the approximation that $\lambda_f \simeq \lambda_i$. Equation 3-41 results.

Problem 3-10. Write

$$\frac{1}{\tau} = \sum_{\text{final states}} \frac{2\pi}{\hbar} |M|^2 \, \delta(E_{2s} - E_{1s} - \hbar c k_1 - \hbar c k_2)$$

$$= \frac{\Omega^2}{(2\pi)^6} \frac{2\pi}{\hbar^2 c} \sum_{\sigma_1} \sum_{\sigma_2} \int d^3k_1 \int d^3k_2 \, |M|^2 \, \delta(k_0 - k_1 - k_2)$$

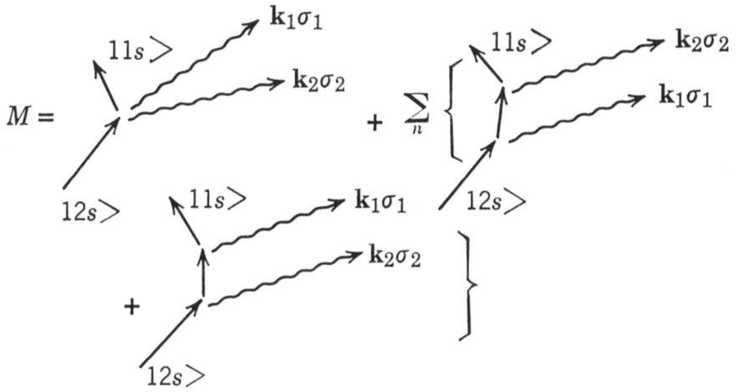

The matrix element for the process can be written as

$$= \frac{e^2}{2mc^2}\left(\frac{2\pi\hbar c^2}{\Omega}\right)\frac{1}{c\sqrt{k_1 k_2}} M'$$

where

$$M' = \mathbf{u}_1 \cdot \mathbf{u}_2 \langle 1s |\, e^{-i(\mathbf{k}_1+\mathbf{k}_2)\cdot\mathbf{x}} \,| 2s\rangle$$

$$+ \frac{2}{m}\sum_n \left\{ \frac{\langle 1s|\, \mathbf{p}\cdot\mathbf{u}_2 \,|n\rangle\langle n|\, \mathbf{p}\cdot\mathbf{u}_1 \,|2s\rangle}{E_{2s} - E_n - \hbar c k_1} + \frac{\langle 1s|\, \mathbf{p}\cdot\mathbf{u}_1 \,|n\rangle\langle n|\, \mathbf{p}\cdot\mathbf{u}_2 \,|2s\rangle}{E_{2s} - E_n - \hbar c k_2} \right\}$$

Note that in the sum over intermediate states it is necessary to include both of the time orders in which the photons are emitted. We get

$$\frac{1}{\tau} = \frac{r_e^2 c}{4(2\pi)^3} \sum_{\sigma_1}\sum_{\sigma_2} \int \frac{d^3k_1}{k_1} \int \frac{d^3k_1}{k_2} |M'|^2 \, \delta(k_0 - k_1 - k_2)$$

Since M' is dimensionless an order of magnitude estimate can be obtained by assuming that

$$\sum_{\sigma_1}\sum_{\sigma_2} |M'|^2 \simeq 1$$

The integrals can now be done and one obtains

$$\frac{1}{\tau} = \frac{1}{12\pi} r_e^2 c k_0^3$$

Using $k_0 = \frac{3}{8}\left(\frac{Z^2}{137}\right)\frac{1}{a}$ we find

$$\frac{1}{\tau} = \frac{Z^6}{12\pi}\left(\frac{3}{8}\right)^3 \frac{1}{(137)^7} \frac{c}{a} = 20Z^6 \text{ sec}^{-1}$$

This problem has attracted considerable attention from theorist beginning with M. Göppert Mayer[63] in 1931. The lastest calculation by Shapiro and Breit[64] in 1959 gives

$$\frac{1}{\tau} = 8.226Z^6 \text{ sec}^{-1}$$

Problem 3-11. Write the conservation of energy as

$$\sqrt{\hbar^2c^2q^2 + m^2c^4} = \sqrt{\hbar^2c^2\,|\mathbf{q} - \mathbf{k}|^2 + m^2c^4} + \frac{\hbar kc}{n}$$

Solving for $\cos\theta = \mathbf{q}\cdot\mathbf{k}/qk$ gives Eq. 3-63.

CHAPTER 4

Problem 4-1. Write

$$[\psi(\mathbf{x}', t), H]_- = \int d^3x \Big\{ \psi(\mathbf{x}', t)\psi^+(\mathbf{x}, t)\Big[-\frac{\hbar^2}{2m}\nabla^2 + V\Big]\psi(\mathbf{x}, t)$$

$$- \psi^+(\mathbf{x}, t)\Big[-\frac{\hbar^2}{2m}\nabla^2 + V\Big]\psi(\mathbf{x}, t)\psi(\mathbf{x}', t)\Big\}$$

Now use Eqs. 4-26 to write $\psi(\mathbf{x}', t)\psi^+(\mathbf{x}, t) = \delta(\mathbf{x} - \mathbf{x}') \pm \psi^+(\mathbf{x}, t)\psi(\mathbf{x}', t)$. Note that $\psi(\mathbf{x}', t)$ can be moved to the right of $[-(\hbar^2/2m)\nabla^2 + V]$ since ∇^2 operates on the unprimed variables. Next use Eq. 4-27 to write

$$\psi(\mathbf{x}', t)\psi(\mathbf{x}, t) = \mp\psi(\mathbf{x}, t)\psi(\mathbf{x}', t)$$

We are left with

$$[\psi(\mathbf{x}', t), H]_- = \int d^3x\ \delta(\mathbf{x} - \mathbf{x}')\Big[-\frac{\hbar^2}{2m}\nabla^2 + V\Big]\psi(\mathbf{x}, t)$$

$$= \Big[-\frac{\hbar^2}{2m}\nabla_{x'}{}^2 + V(\mathbf{x}')\Big]\psi(\mathbf{x}', t)$$

Problem 4-2. Write

$$[N, H]_- = \int d^3x \int d^3x' \Big\{ \psi^+(\mathbf{x}, t)\psi(\mathbf{x}, t)\psi^+(\mathbf{x}', t)\Big[-\frac{\hbar^2}{2m}\nabla_{x'}{}^2 + V\Big]\psi(\mathbf{x}', t)$$

$$- \psi^+(\mathbf{x}', t)\Big[-\frac{\hbar^2}{2m}\nabla_{x'}{}^2 + V\Big]\psi(\mathbf{x}', t)\psi^+(\mathbf{x}, t)\psi(\mathbf{x}, t)\Big\}$$

Use Eqs. 4-26 and 4-27 to move the operator $\psi^+(\mathbf{x}, t)\psi(\mathbf{x}, t)$ in the first term through the operators which stand on its right. Then cancel the resulting expression with the last term. In doing this you pick up a $\delta(\mathbf{x} - \mathbf{x}')$ when $\psi(\mathbf{x}, t)$ is moved through $\psi^+(\mathbf{x}', t)$ and $a - \delta(\mathbf{x} - \mathbf{x}')$ when $\psi^+(x, t)$ is moved through $\psi(\mathbf{x}', t)$, so that these terms cancel.

CHAPTER 5

Problem 5-1

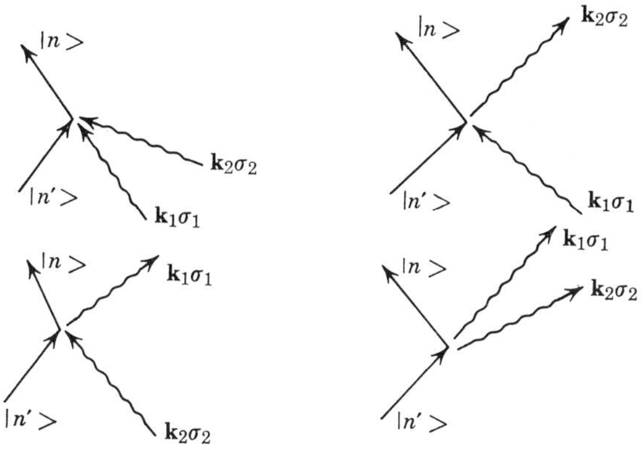

Problem 5-2. In Eq. 5-6b interchange n and n' and note that

$$M(-\mathbf{k}, \sigma, n', n) = M^*(\mathbf{k}, \sigma, n, n')$$

Similar manipulations lead to Eq. 5-8b.

Problem 5-3. No answer necessary.

CHAPTER 6

Problem 6-1. Write

$$\begin{aligned}(\boldsymbol{\alpha} \cdot \mathbf{a})(\boldsymbol{\alpha} \cdot \mathbf{b}) + (\boldsymbol{\alpha} \cdot \mathbf{b})(\boldsymbol{\alpha} \cdot \mathbf{a}) &= 2a_x b_x \alpha_x{}^2 + 2a_y b_y \alpha_y{}^2 + 2a_z b_z \alpha_z{}^2 \\ &\quad + a_x b_y(\alpha_x \alpha_y + \alpha_y \alpha_x) + \cdots\end{aligned}$$

Using Eq. A-24, Eq. 6-30 follows. Next

$$\begin{aligned}\text{Tr}\,(\boldsymbol{\alpha} \cdot \mathbf{a})(\boldsymbol{\alpha} \cdot \mathbf{b}) + \text{Tr}\,(\boldsymbol{\alpha} \cdot \mathbf{b})(\boldsymbol{\alpha} \cdot \mathbf{a}) &= 2\,\text{Tr}\,(\boldsymbol{\alpha} \cdot \mathbf{a})(\boldsymbol{\alpha} \cdot \mathbf{b}) \\ &= 2(\mathbf{a} \cdot \mathbf{b})\,\text{Tr}\,1 = 8\mathbf{a} \cdot \mathbf{b}\end{aligned}$$

from which Eq. 6-32 follows. Note that since β anticommutes with α_i for all i,

$$(\boldsymbol{\alpha} \cdot \mathbf{a})\beta(\boldsymbol{\alpha} \cdot \mathbf{b})\beta = -(\boldsymbol{\alpha} \cdot \mathbf{a})\beta^2(\boldsymbol{\alpha} \cdot \mathbf{b}) = -(\boldsymbol{\alpha} \cdot \mathbf{a})(\boldsymbol{\alpha} \cdot \mathbf{b})$$

from which Eq. 6-33 follows. To prove Eq. 6-34 use Eq. 6-30 to move the factor $(\boldsymbol{\alpha} \cdot \boldsymbol{\alpha})$ to the left; thus

$$\begin{aligned}\operatorname{Tr} (\boldsymbol{\alpha} \cdot \mathbf{a})(\boldsymbol{\alpha} \cdot \mathbf{b})(\boldsymbol{\alpha} \cdot \mathbf{c})(\boldsymbol{\alpha} \cdot \mathbf{d}) = {} & 2(\mathbf{c} \cdot \mathbf{d}) \operatorname{Tr} (\boldsymbol{\alpha} \cdot \mathbf{a})(\boldsymbol{\alpha} \cdot \mathbf{b}) \\ & -2(\mathbf{b} \cdot \mathbf{d}) \operatorname{Tr} (\boldsymbol{\alpha} \cdot \mathbf{a})(\boldsymbol{\alpha} \cdot \mathbf{c}) \\ & + 2(\mathbf{a} \cdot \mathbf{d}) \operatorname{Tr} (\boldsymbol{\alpha} \cdot \mathbf{b})(\boldsymbol{\alpha} \cdot \mathbf{c}) \\ & - \operatorname{Tr} (\boldsymbol{\alpha} \cdot \mathbf{d})(\boldsymbol{\alpha} \cdot \mathbf{a})(\boldsymbol{\alpha} \cdot \mathbf{b})(\boldsymbol{\alpha} \cdot \mathbf{c})\end{aligned}$$

Now, when Eqs. 6-31 and 6-32 are used, Eq. 6-34 is obtained.

Problem 6-2

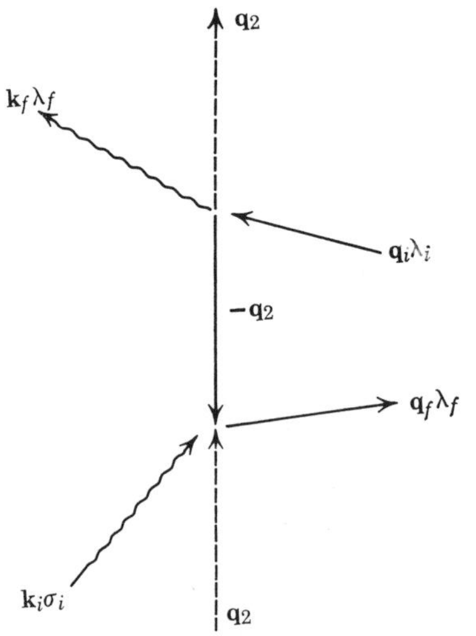

CHAPTER 7

Problem 7-1. Using the definitions $\gamma_5 = \gamma_1\gamma_2\gamma_3\gamma_4$, $\gamma_\mu = -i\beta\alpha_i$, and $\gamma_4 = \beta$ we find

$$\gamma_5 = \begin{pmatrix} 0 & 1 \\ -1 & 0 \end{pmatrix}$$

$$i\gamma_i = \begin{pmatrix} 0 & \sigma_i \\ \sigma_i & 0 \end{pmatrix} \qquad \text{for } i = 1, 2, 3$$

$$i\gamma_4 = i\begin{pmatrix} 1 & 0 \\ 0 & -1 \end{pmatrix}$$

Then

$$i\gamma_i\gamma_5 = \begin{pmatrix} -\sigma_i & 0 \\ 0 & \sigma_i \end{pmatrix} \qquad \text{for } i = 1, 2, 3$$

and

$$i\gamma_4\gamma_5 = \begin{pmatrix} 0 & 1 \\ 1 & 0 \end{pmatrix}$$

All matrices are written as 2 × 2 matrices of 2 × 2 matrices. In the non-relativistic approximation

$$\psi_n = \begin{pmatrix} u_n \\ 0 \end{pmatrix}$$

$$\bar{\psi}_p = i\psi^+\beta = i(u_p^+, 0)\begin{pmatrix} 1 & 0 \\ 0 & -1 \end{pmatrix} = i(u_p^+, 0)$$

where u_n and u_p are 2-component spinors. Equation 7-34 follows.

Problem 7-2. This problem is analogous to Problems 3-5 and 3-6. We may write

$$H_I = -\frac{ie\hbar g}{2Mc}\sum_{\mathbf{k}\sigma}\left(\frac{2\pi\hbar c^2}{\Omega\omega_k}\right)^{\frac{1}{2}} \tau\boldsymbol{\sigma}\cdot(\mathbf{k}\times\mathbf{u}_{\mathbf{k}\sigma})[a_{\mathbf{k}\sigma}e^{i\mathbf{k}\cdot\mathbf{x}} - a_{\mathbf{k}\sigma}^+e^{-i\mathbf{k}\cdot\mathbf{x}}]$$

Assume that the initial and final states are

$$|i\rangle = |\Sigma^0 \uparrow\rangle\,|\text{no photons}\rangle$$
$$|f\rangle = |\Lambda, \uparrow\rangle\,|\cdots 1_{\mathbf{k}}\cdots\rangle,$$

Then

$$\langle f|\, H_I\,|i\rangle = \frac{ie\hbar g}{2Mc}\left(\frac{2\pi\hbar c^2}{\Omega\omega_k}\right)^{\frac{1}{2}}(\mathbf{k}\times\mathbf{u}_{k\sigma})_z$$

The lifetime is given by

$$\frac{1}{\tau} = \frac{\Omega}{(2\pi)^3}\sum_{\sigma}\left(\frac{e\hbar g}{2Mc}\right)^2\left(\frac{2\pi\hbar c^2}{\Omega}\right)\frac{2\pi}{\hbar}\int\frac{d^3k}{kc}(\mathbf{k}\times\mathbf{u}_{k\sigma})_z^2\,\delta[\Delta E - \hbar ck]$$

By comparison with Problem 3-6 we find

$$\frac{1}{\tau} \sim \frac{e^2\hbar}{M^2c^2}k_0^3$$

with

$$k_0 = \frac{1}{\hbar c}\Delta E = \frac{1}{\hbar c}[M_\Sigma c^2 - M_\Lambda c^2]$$
$$= \frac{\Delta Mc}{\hbar}$$

Now $M_\Sigma = 1192\ MeV/c^2$ and $M_\Lambda = 1115\ MeV/c^2$, so that

$$\Delta M = 77\ MeV/c^2$$

We find $\tau \sim 10^{-18}$ sec.

CHAPTER 8

Problem 8-1. When Eq. 8-9 is substituted into Eq. 8-7 it is found that the quantity in braces vanishes whenever the argument of the δ-function vanishes.

Problem 8-2. The proof is almost identical to that of Problem 8-1.

Problem 8-3. When a gas is far from degeneracy, $N(\mathbf{k}) \ll 1$ and Eq. 8-11 reduces to Eq. 8-18.

Problem 8-4. The proof of the classical H-theorem parallels almost exactly the proof outlined in the text for bosons.

Problem 8-5. The critical speed (about 23 cm/sec) is the velocity of propagation of surface waves on the water. Above this, velocity waves propagate away from the moving object. From a quantum-mechanical viewpoint we may say that the moving object can emit hydrons (this is what Synge[65] calls them) and conserve momentum and energy when its velocity exceeds the velocity of propagation of the wave. This is another example of a very general phenomena that have applications to Čerenkov radiation, superfluidity, the wake of a ship, sonic booms, characteristic energy losses of electrons in solids, and Landau damping of plasma oscillations.

CHAPTER 9

Problem 9-1. In the absence of a plasma the potential would be Q/r. The Fourier transform of this is

$$\phi(\mathbf{q}) = \frac{4\pi Q}{\Omega q^2}$$

In the presence of a plasma this must be divided by the dielectric function evaluated at zero frequency (since Q is stationary) to obtain

$$\phi(q) = \frac{4\pi Q}{\Omega q^2 \varepsilon(q, 0)}$$

where

$$\varepsilon(q, 0) = 1 + \frac{1}{q^2 \lambda^2}$$

and

$$\frac{1}{\lambda^2} = \sum_s 3\,\frac{\omega_{ps}^{\ 2}}{V_{fs}^{\ 2}}$$

inverting the Fourier transform gives

$$\begin{aligned}\phi(r) &= \frac{\Omega}{(2\pi)^3}\int d^3q\,\phi(\mathbf{q})e^{i\mathbf{q}\cdot\mathbf{x}}\\ &= \frac{Q}{r}\,e^{-r/\lambda}\end{aligned}$$

The field is shielded out in a distance about equal to λ. A similar result is obtained when the particles have a Maxwellian distribution. In this case

$$\lambda = \left(\frac{T}{4\pi ne^2}\right)^{1/2}$$

is the Debye length.

Problem 9-2. Let

$$f_0(\mathbf{v}) = \frac{N}{\pi^{3/2}\alpha^3}\,e^{-v^2/\alpha^2}$$

where

$$\alpha = \left(\frac{2T}{m}\right)^{1/2}$$

is the thermal velocity. Forget about the sum over species in Eq. 9-24 for the time being. It can be reintroduced later. Take $\mathbf{q}$ along the z-axis and perform the integrations over v_x and v_y to obtain

$$\begin{aligned}\varepsilon(q,\omega) &= 1 - \frac{8\pi ne^2}{mq^2}\,\frac{1}{\alpha^3\sqrt{\pi}}\int_{-\infty}^{+\infty} dv_z\,\frac{qv_z e^{-v_z^{\ 2}/\alpha^2}}{\omega - qv_z}\\ &= 1 + \left(\frac{8\pi ne^2}{m\alpha^2}\right)\frac{1}{q^2} - \left(\frac{8\pi ne^2}{m\alpha^2}\right)\frac{\omega}{q^2\alpha\sqrt{\pi}}\int_{-\infty}^{+\infty} dv_z\,\frac{e^{-v_z^{\ 2}/\alpha^2}}{\omega - qv_z}\\ &= 1 + \frac{1}{\lambda^2 q^2} - \frac{1}{\lambda^2 q^2}\left(\frac{\omega}{q\alpha}\right)Z\left(\frac{\omega}{q\alpha}\right)\end{aligned}$$

where

$$\lambda = \left(\frac{m\alpha^2}{8\pi ne^2}\right)^{1/2} = \left(\frac{T}{4\pi ne^2}\right)^{1/2}$$

and

$$Z(z) = \int_{-\infty}^{+\infty} dx\,\frac{e^{-x^2}}{z - x}$$

is the Fried-Conte function.[66] It has been tabulated.

Problem 9-3. Write

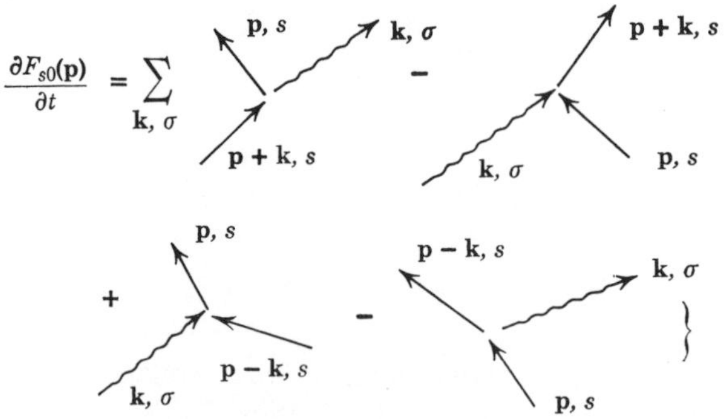

Replacing the diagrams by the corresponding transition probabilities per unit times gives

$$\frac{\partial F_{s0}(\mathbf{p})}{\partial t} = \sum_{\mathbf{k},\sigma} \frac{2\pi}{\hbar} \left| \frac{4\pi e_s{}^2 \hbar \Omega_{k\sigma}}{\Omega k^2 \left(\frac{\partial}{\partial \omega} \omega \varepsilon_1 \right)_{\Omega \mathbf{k}\sigma}} \right|$$

$$\{\delta[E_{s\mathbf{p}+\mathbf{k}} - E_{s\mathbf{p}} - \hbar\Omega_{\mathbf{k}\sigma}](F_{s0}(\mathbf{p}+\mathbf{k})[1 \pm F_{s0}(\mathbf{p})]$$
$$[N_\sigma(\mathbf{k}) + 1] - F_{s0}(\mathbf{p})[1 \pm F_{s0}(\mathbf{p}+\mathbf{k})]N_\sigma(\mathbf{k}))$$
$$+ \delta[E_{s\mathbf{p}-\mathbf{k}} - E_{sp} + \hbar\Omega_{\mathbf{k}\sigma}](F_{s0}(\mathbf{p}-\mathbf{k})[1 \pm F_{s0}(\mathbf{p})]$$
$$N_\sigma(\mathbf{k}) - F_{s0}(\mathbf{p})[1 \pm F_{s0}(\mathbf{p}-\mathbf{k})][N_\sigma(\mathbf{k}) + 1])\}$$

The quasi-linear equations were first derived classically by Drummond and Pines[67] and by Vedenov, Velikov, and Sagdeev.[68] The quantum-mechanical derivation is due to Pines and Schrieffer.[69]

CHAPTER 10

Problem 10-1. Solving

$$\frac{d^2}{dt^2} \Delta \mathbf{x} = \frac{e}{m} \mathbf{E}$$

on the assumption the $\mathbf{E}$ oscillates with frequency ω gives

$$\Delta \mathbf{x} = -\frac{e}{m\omega^2} \mathbf{E}$$

and

$$|\Delta \mathbf{x}|^2 = \frac{e^2}{m^2 \omega^4} E^2$$

Now, we let E be one of the modes of the radiation field and obtain

$$\langle|\Delta\mathbf{x}|^2\rangle = \frac{e^2}{m^2}\sum_{\mathbf{k},\sigma}\frac{1}{\omega_k^{\ 4}}E_{\mathbf{k}\sigma}^{\ 2}$$

We use

$$\frac{E_{\mathbf{k}\sigma}^{\ 2}}{4\pi} = \frac{\hbar\omega_k}{2\Omega}$$

and obtain

$$\begin{aligned}\langle|\Delta\mathbf{x}|^2\rangle &= \frac{2\pi\hbar}{\Omega}\frac{e^2}{m^2}\sum_{\mathbf{k},\sigma}\frac{1}{\omega_k^{\ 3}}\\ &= \frac{2\pi\hbar}{\Omega}\frac{e^2}{m^2}\frac{\Omega}{(2\pi)^3}2\int d^3k\,\frac{1}{(kc)^3}\\ &= \frac{2}{\pi}\frac{e^2}{\hbar c}\left(\frac{\hbar}{mc}\right)^2\int_0^\infty\frac{dk}{k}\end{aligned}$$

When the sum over field modes is made the term $\mathbf{\Delta x}\cdot\mathbf{\nabla}$ and the off-diagonal terms in $\frac{1}{2}(\mathbf{\Delta}x\cdot\nabla)^2$ vanish. One is left with Eq. 10-58. Letting

$$V = -\frac{e^2}{r}$$

we find

$$\tfrac{1}{6}\langle|\mathbf{\Delta}x|^2\rangle\,\nabla^2 V = -\frac{2\pi e^2}{3}\langle|\Delta\mathbf{x}|^2\rangle\,\delta(\mathbf{x}) = \frac{4}{3}\frac{e^4}{\hbar c}\left(\frac{\hbar}{mc}\right)^2\log\left(\frac{k_{\max}}{k_{\min}}\right)\delta(\mathbf{x})$$

This gives an energy shift of

$$\Delta E = \frac{4}{3}\frac{e^2}{\hbar c}\left(\frac{\hbar}{mc}\right)^2\log\left(\frac{k_{\max}}{k_{\min}}\right)|\psi(0)|^2$$

Which agrees with Eq. 10-33 if the cutoffs $k_{\max}$ and $k_{\min}$ are properly chosen. The reader is referred to Welton's paper for the arguments justifying the choice of $k_{\max}$ and $k_{\min}$.

APPENDIX A

Problem A-1. Multiply Eq. A-5 by $a_{\mu\gamma}$ to obtain

$$a_{\mu\gamma}x'_\mu = a_{\mu\gamma}a_{\mu\lambda}x_\lambda = \delta_{\gamma\lambda}x_\lambda = x_\gamma$$

Now use Eq. A-18 twice in $x_\lambda x_\lambda = x'_\mu x'_\mu$ to obtain

$$a_{\mu\lambda}a_{\beta\lambda}x'_\mu x'_\beta = x'_\mu x'_\mu$$

If this is to be true for all x'_μ it follows that $a_{\mu\lambda}a_{\beta\lambda} = \delta_{\mu\beta}$. By the chain rule of differentiation

$$\frac{\partial}{\partial x_\mu} = \frac{\partial x'_\lambda}{\partial x_\mu}\frac{\partial}{\partial x'_\lambda} = a_{\lambda\mu}\frac{\partial}{\partial x'_\lambda}$$

Similarly

$$\frac{\partial^2}{\partial x_\mu\,\partial x_\mu} = a_{\lambda\mu}a_{\beta\mu}\frac{\partial^2}{\partial x'_\lambda\,\partial x'_\beta} = \delta_{\lambda\beta}\frac{\partial^2}{\partial x'_\lambda\,\partial x'_\beta} = \frac{\partial^2}{\partial x'_\lambda\,\partial x'_\lambda}$$

Problem A-2. Write

$$\begin{aligned} a_{\mu\nu}a_{\mu\lambda} &= (\delta_{\mu\nu} + \varepsilon_{\mu\nu})(\delta_{\mu\lambda} + \varepsilon_{\mu\nu}) \\ &= \delta_{\nu\lambda} + \varepsilon_{\lambda\nu} + \varepsilon_{\nu\lambda} + \varepsilon_{\mu\nu}\varepsilon_{\mu\lambda} = \delta_{\mu\lambda} \end{aligned}$$

Since $\varepsilon_{\nu\lambda}$ is infinitesimal the term $\varepsilon_{\mu\nu}\varepsilon_{\mu\lambda}$ is negligible and it follows that $\varepsilon_{\nu\lambda} = -\varepsilon_{\lambda\nu}$. Equation A-36 becomes

$$(1 - T)\lambda_\mu(1 + T) = (\lambda_{\mu\lambda} + \varepsilon_{\mu\lambda})\gamma_\lambda$$

from which

$$\gamma_\mu T - T\gamma_\mu = \varepsilon_{\mu\lambda}\gamma_\lambda$$

This is easily seen to be satisfied by

$$T = \tfrac{1}{4}\varepsilon_{\mu\nu}\gamma_\mu\gamma_\nu$$

when $\gamma_\mu\gamma_\nu + \gamma_\nu\gamma_\mu = 2\delta_{\mu\nu}$ is used.

The Lorentz transformation corresponding to a rotation through an angle ϕ about the z-axis is given by

$$\begin{aligned} x'_1 &= x_1 \cos\phi + x_2 \sin\phi \\ x'_2 &= -x_1 \sin\phi + x_2 \cos\phi \\ x'_3 &= x_3 \\ x'_4 &= x_4 \end{aligned}$$

from which

$$\begin{aligned} a_{33} &= a_{44} = 1 \\ a_{11} &= a_{22} = \cos\phi \\ a_{12} &= -a_{21} = +\sin\phi \end{aligned}$$

and all others are zero. For a rotation through an infinitesimal angle ε we have

$$\varepsilon_{12} = -\varepsilon_{21} = +\varepsilon$$

so

$$\begin{aligned} S_\varepsilon &= 1 + \tfrac{1}{4}\varepsilon_{12}\gamma_1\gamma_2 + \tfrac{1}{4}\varepsilon_{21}\gamma_2\gamma_1 \\ &= 1 + \frac{\varepsilon}{2}\gamma_1\gamma_2 \end{aligned}$$

We can find S for a rotation through a finite angle ϕ by iterating this the proper number of times. That is

$$S_\phi = \left(1 + \frac{\varepsilon}{2}\gamma_1\gamma_2\right)^{\phi/\varepsilon}$$

In the limit $\varepsilon \to 0$ this becomes

$$S_\phi = e^{(\phi/2)\gamma_1\gamma_2}$$

By direct calculation one finds

$$\gamma_1\gamma_2 = \begin{pmatrix} +i & 0 & 0 & 0 \\ 0 & -i & 0 & 0 \\ 0 & 0 & +i & 0 \\ 0 & 0 & 0 & -i \end{pmatrix}$$

and so S_ϕ is given by Eq. A-46.

For the next part of the problem it is useful to consider the finite Lorentz transformation as a rotation in the $x_1 - x_4$-plane through an imaginary angle ϕ such that

$$\cos\phi = \frac{1}{\sqrt{1-\beta^2}}, \qquad \sin\phi = \frac{i\beta}{\sqrt{1-\beta^2}}$$

where $\beta = v/c$. If $\phi \to \delta$ where δ is infinitesimal, then

$$S_\delta = 1 + \frac{\delta}{2}\gamma_1\gamma_4 = 1 + \frac{i\delta}{2}\alpha_1$$

Then

$$\begin{aligned} S_\phi &= \left(1 + \frac{i\delta}{2}\alpha_1\right)^{\phi/\delta} = e^{i(\phi/2)\alpha_1} \\ &= \sum_{n=0}^{\infty}\frac{1}{n!}\left(\frac{i\phi}{2}\right)^n \alpha_1{}^n \\ &= 1\cos\frac{\phi}{2} + \alpha_1 \sin\frac{\phi}{2} \\ &= 1\sqrt{\frac{1+\cos\phi}{2}} + i\alpha_1\sqrt{\frac{1-\cos\phi}{2}} \\ &= 1\sqrt{\frac{1+\sqrt{1-\beta^2}}{2\sqrt{1-\beta^2}}} + \alpha_1\sqrt{\frac{1-\sqrt{1-\beta^2}}{2\sqrt{1-\beta^2}}} \end{aligned}$$

Problem A-3. For the space reflection transformation

$$a_{11} = a_{22} = a_{33} = -1 \qquad \text{and} \qquad a_{44} = +1$$

Equation A-36 gives

$$S^{-1}\gamma_1 S = -\gamma_1$$
$$S^{-1}\gamma_2 S = -\gamma_2$$
$$S^{-1}\gamma_3 S = -\gamma_3$$
$$S^{-1}\gamma_4 S = +\gamma_4$$

These equations are clearly satisfied by $S = \gamma_4$, since

$$\gamma_4\gamma_i = -\gamma_i\gamma_4 \qquad \text{for } i = 1, 2, 3$$

and

$$\gamma_4^{\ 2} = 1$$

Problem A-4. Consider

$$L_z = xp_y - yp_x$$

The lack of commutativity must come from the terms in H containing p_x and p_y. We find

$$[L_z, p_x]_- = -\frac{\hbar}{i}\, p_y$$

$$[L_z, p_y]_- = +\frac{\hbar}{i}\, p_x$$

so

$$[L_z, H]_- = -\frac{\hbar c}{i}\,\alpha_x p_y + \frac{\hbar c}{i}\,\alpha_y p_x$$

Now consider

$$S_z = \frac{\hbar}{2i}\,\alpha_x\alpha_y$$

We find

$$[S_z, \alpha_x]_- = -\frac{\hbar}{i}\,\alpha_y$$

$$[S_z, \alpha_y] = \frac{\hbar}{i}\,\alpha_x$$

$$[S_z, \alpha_z] = 0$$

so

$$[S_z, H] = -\hbar c\alpha_y p_x + \hbar c\alpha_x p_y = -[L_z, H]$$

When L_z and S_z are added to obtain J_z, their sum commutes with H. A similar proof holds for the other components.

One can show that

$$\mathbf{S} = \frac{\hbar}{2}\begin{pmatrix} \boldsymbol{\sigma} & 0 \\ 0 & \boldsymbol{\sigma} \end{pmatrix}$$

It follows that the eigenvalues of the components of **S** are $\hbar/2$ and the eigenvalues of S^2 are $3\hbar^2/4$.

Problem A-5. Using Eq. A-25 for $\boldsymbol{\alpha}$ and β and making the usual replacement $\mathbf{p} \to \mathbf{p} - e/c\mathbf{A}$ and $E \to E - e\Phi$ it is straightforward to derive Eqs. A-61. In the nonrelativistic limit we have $\chi \ll \phi$ for a positive energy electron. Furthermore,

$$\begin{bmatrix} \phi \\ \chi \end{bmatrix} \sim e^{-i/\hbar E t}$$

and $E \simeq mc^2$, so that Eq. A-61b gives

$$\chi \simeq \frac{1}{2mc}\left(\frac{\hbar}{i}\nabla - \frac{e}{c}\mathbf{A}\right) \cdot \boldsymbol{\sigma}\phi$$

Next we can show that

$$(\mathbf{a} \cdot \boldsymbol{\sigma})(\mathbf{b} \cdot \boldsymbol{\sigma}) = \mathbf{a} \cdot \mathbf{b} + i\boldsymbol{\sigma} \cdot (\mathbf{a} \times \mathbf{b})$$

when **a** and **b** are noncommuting operators. Applying this to

$$\mathbf{a} = \mathbf{b} = \left(\frac{\hbar}{i}\nabla - \frac{e}{c}\mathbf{A}\right)$$

gives

$$\left|\left(\frac{\hbar}{i}\nabla - \frac{e}{c}\mathbf{A}\right) \cdot \boldsymbol{\sigma}\right|^2 = \left(\frac{\hbar}{i}\nabla - \frac{e}{c}\mathbf{A}\right)^2 - \frac{e\hbar}{c}(\nabla \times \mathbf{A}) \cdot \boldsymbol{\sigma}$$

Equation A-62 results.

Problem A-6. Write

$$H\psi = -c\begin{bmatrix} p_z & p_- \\ p_+ & -p_z \end{bmatrix}\begin{pmatrix} \psi_1 \\ \psi_2 \end{pmatrix} = E\begin{pmatrix} \psi_1 \\ \psi_2 \end{pmatrix}$$

Where as usual $p_\pm = p_x \pm ip_y$. Equating the determinant of the coefficients to zero and solving for E gives

$$E^2 - c^2p_z^2 - c^2p_+p_- = E^2 - c^2p^2 = 0$$

from which $E = \pm cp$. The properly normalized eigenfunctions are

$$\psi_+ = \begin{bmatrix} \sqrt{\dfrac{p - p_z}{2p}} \\ -\dfrac{p_+}{\sqrt{2p(p - p_z)}} \end{bmatrix}$$

and

$$\psi_- = \begin{bmatrix} \sqrt{\dfrac{p + p_z}{2p}} \\ -\dfrac{p_+}{\sqrt{2p(p + p_z)}} \end{bmatrix}$$

The proof of part (*b*) is almost identical to that of Problem A-4. Writing

$$\sigma_{\mathbf{p}} = \frac{\mathbf{p} \cdot \boldsymbol{\sigma}}{p}$$

For the spin operator in the direction of the momentum gives

$$H = -E\sigma_{\mathbf{p}}$$

It is obvious that H and $\sigma_{\mathbf{p}}$ have the same eigenfunctions. The eigenfunction corresponding to the eigenvalue $E = +cp$ of H has the eigenvalue -1 of $\sigma_{\mathbf{p}}$ and the eigenvalue $-\hbar/2$ of $(\hbar/2)\sigma_{\mathbf{p}}$.

Notes and References

CHAPTER 1

Quantum mechanics began with two quite different mathematical formulations: the differential equation of Schrödinger[1] and the matrix algebra of Heisenberg.[2] These two points of view were ultimately synthesized in the transformation theory of Dirac.[3] A third formulation, the space-time approach of Feynman,[4] has played an important role in quantum-field theory, but it is not used in this book. The theory of Hilbert space was introduced into quantum mechanics by von Neumann[5] in order to establish a rigorous mathematical basis for the theory. A good modern account of the foundations of quantum mechanics is the book of Jauch.[6] In writing this chapter I have borrowed from the very concise exposition of Sen.[7]

1. E. Schrödinger, *Ann. Physik* **79,** 361, 489, 734; **80,** 437; **81,** 109 (1926).
2. W. Heisenberg, *Z. Phys.* **33,** 879 (1925).
3. P. A. M. Dirac, *Principles of Quantum Mechanics*, Clarendon Press, Oxford, 1930; 4th ed., 1958.
4. R. P. Feynman, *Rev. Mod. Phys.* **20,** 267 (1948); Richard P. Feynman and Albert R. Hibbs, *Quantum Mechanics and Path Integrals*, McGraw-Hill, New York, 1965.
5. J. von Neumann, *Mathematische Grundlagen der Quantenmechanik*, Springer, Berlin, 1932; English translation: Princeton University Press, 1955.
6. J. M. Jauch, *Foundations of Quantum Mechanics*, Addison-Wesley, Reading, Mass., 1968.
7. D. K. Sen, *Fields and/or Particles*, Academic Press, London and New York, 1968.

CHAPTERS 2 AND 3

A collection of some of the fundamental papers in quantum electrodynamics, edited by Schwinger, has been published.[8] Dirac[9] was the first to treat the emission and absorption of radiation quantum mechanically. The method we followed in Chapters 2 and 3 is essentially that in Fermi's very readable review article[10] and in the first edition of Heitler's book.[11] A good introduction to noncovariant quantum electrodynamics is the brief book by Power.[12] The theory of coherent states of the radiation field due to Glauber[13] is relatively new. The quantum theory of Čerenkov radiation is due to Ginsberg.[14]

8. *Selected Papers on Quantum Electrodynamics*, Julian Schwinger, Ed., Dover, New York, 1958.
9. P. A. M. Dirac, *Proc. Roy. Soc* **A114,** 243 (1927).
10. E. Fermi, *Rev. Mod. Phys.* **4,** 87 (1932).
11. W. Heitler, *The Quantum Theory of Radiation*, Clarendon Press, Oxford, 1936.
12. E. A. Power, *Introductory Quantum Electrodynamics*, Longmans, Green and Co., London, 1964.
13. Roy J. Glauber, *Phys. Rev. Lett.* **10,** 84 (1963).
14. V. L. Ginsberg, *J. Phys.* (*U.S.S.R.*) **11,** 441 (1940).
15. L. D. Landau and E. M. Lifshitz, *Electrodynamics of Continuous Media* Addison-Wesley, Reading, Mass., 1960.
16. V. Weisskopf and E. Wigner, *Z. Phys.* **63,** 54; **65,** 18 (1930).

CHAPTERS 4 AND 5

The second quantization formalism originated with Jordan, Klein, and Wigner.[17,18] The first part of the book by Henley and Thirring[19] contains a useful and readable introduction to quantum field theory. The bremstrahlung cross section has been discussed by Bethe and Salpeter.[20]

17. P. Jordan and O. Klein, *Z.* Phys. **45,** 751 (1927).
18. P. Jordan and E. P. Wigner, *Z. Phys.* **47,** 631 (1928).
19. Ernest M. Henley and Walter Thirring, *Elementary Quantum Field Theory*, McGraw-Hill, New York, 1962.
20. H. A. Bethe and E. E. Salpeter, *Quantum Theory of One and Two Electron Atoms*, Academic Press, New York, 1957.

CHAPTER 6

There is a large literature on quantum electrodynamics. I have already mentioned the collection of original papers by Schwinger,[8] the review by Fermi,[10] and the book by Power.[12] Some modern textbooks on the subject are those of Heitler,[21] Jauch and Rohrlich,[22] Achieser and Berestetski,[23] Bjorken and Drell,[24] and Thirring,[25] The books of Wentzel,[26] Schweber,[27] and Bogoliubov and Shirkov[28] are general works on quantum-field theory including quantum electrodynamics. Two books by Feynman[29,30] are notable for their informal style and readability.

21. W. Heitler, *The Quantum Theory of Radiation*, 3rd ed., Clarendon Press, Oxford, 1954.
22. J. M. Jauch and F. Rohrlich, *Theory of Photons and Electrons*, Addison-Wesley, Cambridge, 1955.
23. A. I. Achieser and V. B. Berestetski, *Quantum Electrodynamics*, 2nd ed., Wiley, New York, 1963.
24. J. D. Bjorken and S. D. Drell, *Relativistic Quantum Mechanics*, McGraw-Hill, New York, 1964.
25. W. E. Thirring, *Principles of Quantum Electrodynamics*, Academic Press, New York, 1958.
26. G. Wentzel, *Quantum Theory of Fields*, Translated by J. M. Jauch, Interscience, New York, 1949.
27. S. S. Schweber, *An Introduction to Relativistic Quantum Field Theory*, Harper and Row, New York, 1962.
28. N. N. Bogoliubov and D. V. Shirkov, *Introduction to the Theory of Quantized Fields*, Interscience, New York, 1959.
29. R. P. Feynman, *Quantum Electrodynamics*, W. A. Benjamin, New York, 1961.
30. R. P. Feynman, *Theory of Fundamental Processes*, W. A. Benjamin, New York, 1962.
31. J. R. Oppenheimer, *Phys. Rev.* **35,** 939 (1930).
32. P. A. M. Dirac, *Proc. Camb. Phil. Soc.* **26,** 361 (1930).
33. I. Tamm, *Z. Phys.* **62,** 7 (1930).
34. O. Klein and Y. Nishima, *Z. Phys.* **52,** 853 (1929); Y. Nishima, *ibid.* **52,** 869 (1929). The same formula has also been derived by I. Tamm, *ibid.* **62,** 545 (1930).

CHAPTER 7

An interesting account of the origin of Fermi's theory of beta-decay has been given by Rasetti.[35] Fermi originally intended to announce the results of his beta-decay theory in a letter to *Nature*, but this manuscript was rejected by the editor of that journal as containing abstract speculation too remote from physical reality to be of interest to the readers. Fermi then sent a somewhat longer paper to *Ricerca Scientifica* where it was promptly published.[36] He then published full accounts of his theory in both Italian[37] and German.[38] Fortunately for those who read only English, this valuable paper has been translated by Wilson.[39] Modern accounts of the theory of the beta-decay interaction will be found in the books of Feynman,[30] Gasiorowicz,[40] Källén,[51] and Bernstein.[42]

35. Enrico Fermi, *Collected Papers*, University of Chicago Press, Chicago, 1962, p. 538.
36. E. Fermi, *Ric. Sci.* **4,** 491 (1933).
37. E. Fermi, *Nuovo Cimento* **11,** 1 (1934).

38. E. Fermi, *Z. Phys.* **88,** 161 (1934).
39. Fred L. Wilson, *Am. J. Phys.* **36,** 1150 (1968).
40. Stephen Gasiorowicz, *Elementary Particle Physics*, Wiley, New York, 1966.
41. Gunnar Källén, *Elementary Particle Physics*, Addison-Wesley, Reading, Mass., 1964.
42. Jeremy Bernstein, *Elementary Particles and Their Currents*, W. H. Freeman, San Francisco, 1968.

CHAPTER 8 AND 9

There are now a number of books devoted to the application of the methods of quantum-field theory to statistical mechanics and the many body problem[43–45] solid state physics,[46] and plasma physics.[49,50]

43. A. A. Abrikosov, L. P. Gorkov, and I. E. Dzyaloshinski, *Methods of Quantum Field Theory in Statistical Physics*, Prentice-Hall, Englewood Cliffs, N.J., 1963.
44. Leo P. Kadanoff and Gordon Baym, *Quantum Statistical Mechanics*, W. A. Benjamin, New York, 1962.
45. D. Pines, *The Many Body Problem*, W. A. Benjamin, New York, 1961.
46. David Pines, *Elementary Excitations in Solids*, W. A. Benjamin, New York, 1963.
47. P. W. Anderson, *Concepts in Solids*, W. A. Benjamin, New York, 1963.
48. C. Kittel, *Quantum Theory of Solids*, Wiley, New York, 1964.
49. D. F. DuBois, in *Lectures in Theoretical Physics* IXC, W. E. Britten, Ed., Gordon and Breach, 1967.
50. Edward G. Harris, in *Advances in Plasma Physics*, Vol. 3, A. Simon and W. B. Thompson, Eds., Wiley-Interscience, New York, 1969.
51. L. D. Landau and E. M. Lifschitz, *Statistical Physics*, Pergamon Press, London, 1958.
52. L. D. Landau, *J. Phys.* (*U.S.S.R.*) **10,** 25 (1946).

CHAPTER 10

The problem of infinities in quantum electrodynamics is discussed in references 21, 22, 23, 24, 25, 26, 27, 28, and 29. The theory of the attractive force between metal surfaces is due to Casimir[53] and Lifschitz.[54] The experimental observations have been reported by Deryagin and Abrikosava.[55] The radiative correction to the magnetic moment of the electron was first calculated by Schwinger.[60] A very clear treatment of this problem has been given by Luttinger.[61] Arunasalam[62] has given a nonrelativistic theory of this effect.

53. H. B. G. Casimir, *Proc. Nederlands Aka. Wettenshappen, Amsterdam* **60,** 793 (1948).
54. E. M. Lifschitz, *Soviet Phys. JETP* **2,** 73 (1956).

55. B. V. Deryagin and I. I. Abrikosava, *Soviet Phys. JETP* **3,** 819 (1957); **4,** 2 (1957).
56. P. A. M. Dirac, *Phys. Rev.* **139,** B684 (1965).
57. W. E. Lamb, Jr., and R. C. Retherford, *Phys. Rev.* **72,** 241 (1947).
58. H. A. Bethe, *Phys. Rev.* **72,** 339 (1947).
59. Norman M. Kroll and Willis E. Lamb, Jr., *Phys. Rev.* **75,** 388 (1949).
60. Julian Schwinger, *Phys. Rev.* **73,** 416 (1948).
61. J. M. Luttinger, *Phys. Rev.* **74,** 893 (1948).
62. V. Arunasalam, *Am. J. Phys.* **37,** 877 (1969).

APPENDICES

63. M. Göppert-Mayer, *Ann. Phys.* **9,** 273 (1931).
64. J. Shapiro and G. Breit, *Phys. Rev.* **113,** 179 (1959).
65. J. L. Synge, *Science* **138,** 13 (1962).
66. B. D. Fried and C. Conte, *The Plasma Dispersion Function*, Academic Press, New York, 1961.
67. W. E. Drummond and D. Pines, *Nucl. Fusion Suppl.*, Pt. 3, 1049 (1962).
68. A. A. Vedenov, E. P. Velikhov, and R. Z. Sagdeev, *Nucl. Fusion Suppl.*, Pt. 2, 465 (1962).
69. D. Pines and J. R. Schrieffer, *Phys. Rev.* **125,** 804 (1962).
70. T. A. Welton, *Phys. Rev.* **74,** 1157 (1948).

Index